DATE DUE

Demco, Inc. 38-293

Oil Spills

Look for these and other books in the Lucent Overview Series:

Acid Rain

Endangered Species

Energy Alternatives

Garbage

The Greenhouse Effect

Hazardous Waste

Ocean Pollution

Oil Spills

Ozone

Pesticides

Rainforests

Recycling

Saving the American Wilderness

Vanishing Wetlands

Zoos

Oil Spills

by Lesley A. DuTemple

Library of Congress Cataloging-in-Publication Data

DuTemple, Lesley A.
Oil Spills / by Lesley A. DuTemple.
p. cm. — (Lucent overview series. Our endangered planet)
Includes bibliographical references and index.
Summary: Examines the causes and some specific occurrences of oil spills and leaks and discusses the environmental issues and efforts to prevent future spills.
ISBN 1-56006-524-9
1. Oils spills—Environmental aspects Juvenile literature.
[1. Oil spills. 2. Pollution.] I. Title. II. Series.
TD427.P4D88 1999
363.738'2—dc21 99-21614
CIP

P.O. Box 289011, San Diego, CA 92198-9011
Printed in the U.S.A.

Contents

Introduction

OIL LEAKS AND spills into our environment on a daily basis. The public hears about huge oil spills—such as when tankers smash into rocks, break apart in rough seas, run aground in shallow water, or even crash into each other—but smaller spills happen constantly. Train cars derail, trucks skid and crash, jets spill oil during refueling maneuvers, some cruise ships illegally pump used oil overboard, and cars leak oil onto driveways. Drop by drop, barrel by barrel, the spills add up.

The amount of oil that spills into our environment is staggering. So much oil spills that even some enormous spills, such as the *Exxon Valdez*, actually comprise a small fraction of the total amount of oil spilled.

People remember the 1989 *Exxon Valdez* oil spill in Alaska's Prince William Sound and the damage that 258,000 barrels of spilled oil wreaked on the environment. Although the *Exxon Valdez* was the worst spill that had ever occurred in U.S. waters, it wasn't the largest by any measure. Of all major oil spills in the world, it ranks twenty-ninth for amount of oil spilled. Despite its ranking, many people remember this as a horrendous oil spill.

Prevention is key

As more oil is transported around the globe daily, the potential for more deadly and costly oil spills increases. No oil spill can be completely cleaned up, no matter how small it is. Some damage is always done and some portion of a precious natural resource is always lost. To minimize

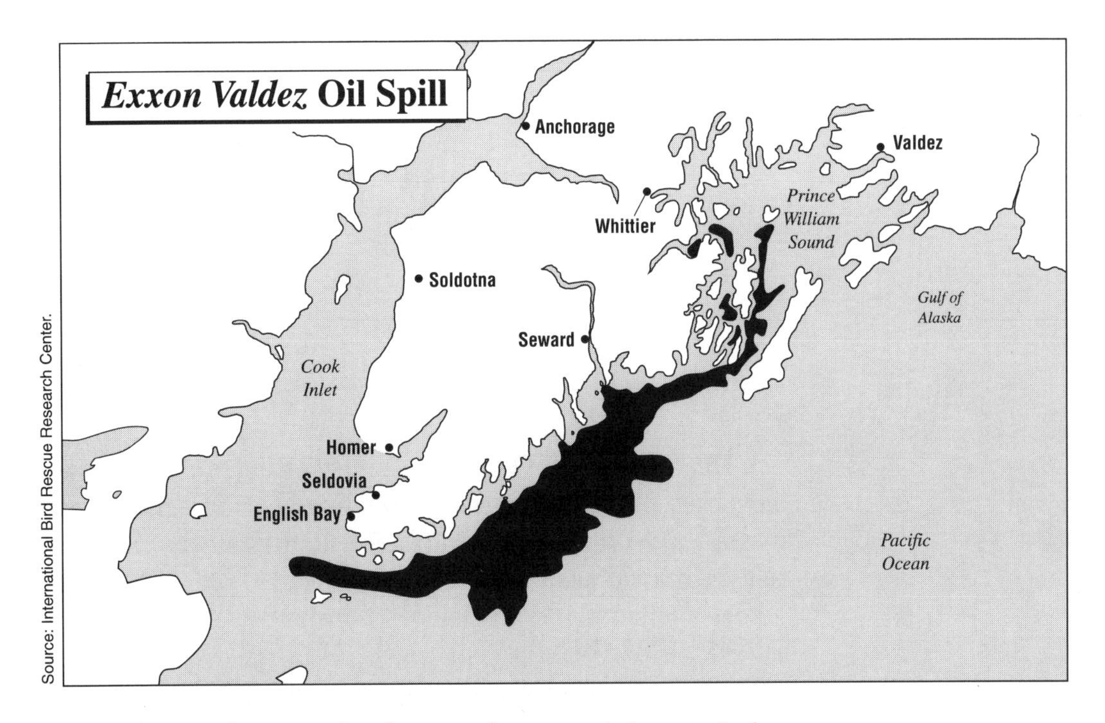

these losses, better technology and crew training are being developed for tankers, and new cleanup methods are also being developed. Ultimately, the only way to prevent this damage and loss is to prevent spills from ever happening in the first place.

1

Oil from Rocks

THE CORRECT NAME for oil is petroleum. The word *petroleum* comes from two Latin words: *petra,* meaning "rock," and *oleum,* meaning "oil." Petroleum is exactly that: oil that comes from rocks.

Ocean beginnings

The majority of the world's petroleum was formed more than a million years ago, during a time when much of the earth's surface was covered by a warm, shallow sea that teemed with life. Millions of plants and animals thrived in this environment, and they reproduced abundantly in the warm, salty water.

As the organisms died, their remains sank to the bottom of the sea. Over the centuries thick layers of organic debris accumulated on the seafloor. Mud mixed with the debris, and sand covered the organic layers. Layer by layer, debris built up on the seafloor. As more debris sank and was added to the top, the bottom layers of debris were compacted and pushed deeper into the earth's crust. The pressure caused the layers of debris to form into layers of sedimentary rock. According to petroleum geologist Norman Hyne, "These ancient sediments, piled layer upon layer, form the sedimentary rocks that make up the uppermost part of the crust of the earth in petroleum-bearing areas."[1] Trapped within the sedimentary rock layers, sometimes in open pockets, sometimes mixed with grains of sand, the plants compressed into the thick, gooey substance that we call petroleum.

All petroleum forms in the same manner: organic matter, sediment layers, and pressure. Pressure is what transforms the organic matter and sedimentary layers into hydrocarbons, or compounds of hydrogen and carbon. Hydrocarbons are the very basis of petroleum, a testament to its organic origins.

Oil reserves around the world

Oil is found all over the globe. Alaska, California, Oklahoma, and Texas have the largest oil reserves in the United States. The former Soviet Union has large oil reserves, as does South America. There is also oil beneath the North Sea, which is mined by both Norway and Britain. Most of the world's oil, however, comes from the Middle East.

The Trans-Alaska Pipeline. Alaska has one of the largest oil reserves in the United States.

Most oil is found deep underground, trapped within and between layers of sedimentary rock. Sandstone, limestone, and shale are all forms of sedimentary rock. Some sandstone deposits can be several miles long and hundreds of feet thick. The sandstone acts like a giant sponge, trapping the oil between the porous grains of sand. Oil deposits can also be found trapped between layers of salt.

Ancient uses of petroleum

Humans have been using liquid petroleum for more than six thousand years. Ancient peoples probably first found petroleum seeping from rocks or coating the water of marshes with a greasy rainbow-colored film. It wasn't long before they figured out ways to use it.

According to a fact sheet published by the Chevron Corporation, "Pitch, a thick form of oil, is referred to in the Bible as the substance used to caulk Noah's ark and baby Moses' basket. American Indians used pitch to waterproof canoes and to make warpaint and medicines."[2] Asphalt, or tar, was often used by ancient peoples as mortar in the walls of their dwellings. Likewise, ancient Egyptians used tar to hold jewels in their settings and to embalm mummies. Even the walls of the pyramids were held together with sticky petroleum.

People also quickly discovered the potential of petroleum as a fuel and for other purposes as well. Liquid petroleum was used as a weapon: Petroleum-filled trenches were set on fire and used as a defense by ancient cities. Other times petroleum was poured on water (an ancient, deliberate oil spill) and set on fire, keeping enemy ships at a distance or destroying them.

The oil derrick has greatly improved modern drilling techniques, and new oil reserves are tapped every day to meet the ever-growing demands of our society.

Drilling for oil

Very little oil is just sitting on the surface of the earth or is easily obtainable. Today petroleum geologists, who are usually hired by large oil companies, seek out places where oil would likely be—such as sedimentary rock deposits—and drill for it. Often, they may need to drill as much as two miles deep.

To drill for oil, a large towerlike structure called a derrick is constructed over a suspected oil deposit. The derrick, or rig, holds the drill and the machinery that will power it. The drill then creates a shaft through dirt and rock in an attempt to reach the underground petroleum reservoirs.

Today virtually all oil is obtained by drilling, either on land or on the seabed. Petroleum geologists are constantly scouting out areas around the globe, looking for new places to drill for oil. New technology is helping these geologists discover reserves in unexpected places, and improved equipment is allowing drillers to tap reserves that were previously inaccessible.

In an effort to satisfy the world's need for oil, more and more wells are being drilled. According to a March 1998 article in the *Houston News Today,* "The Texas Railroad

Commission (the organization in charge of oil wells in Texas) issued a total of 979 original drilling permits."[3] And that's just in Texas alone, and only in one month. In the United States, more than 3 million wells have been drilled. According to the American Petroleum Institute, 6,876 wells were drilled in the United States in the second quarter of 1998.

Types of petroleum

Oil comes out of the ground in several forms: tar, asphalt, bitumen, and crude. Primarily, it comes out as crude oil. Crude oil is also known as liquid petroleum, the same oil that ancient peoples obtained from seepages. Crude oil is petroleum in its most natural state; it burns well, but that's about all.

Refining crude

Once crude oil is extracted from the ground, it needs to be refined. No type of crude oil is usable until it has been refined.

The first step in the refining process is to heat the crude oil until it begins to boil. As the oil is heated, various components of it, called fractions, come to a boil at different temperatures. Like all liquids, when a fraction comes to a boil it produces gas. At refineries, these gases are trapped, cooled, and turned back into liquids.

The first fraction to boil off at a refinery is gasoline. As the temperature increases, kerosene is the next fraction that boils off. Finally, diesel and heating oils boil off at the highest temperatures. When the refining process is complete, only a gooey residue remains, which is used to make the asphalt used for paving roads.

Some of the less valuable petroleum fractions are further refined into chemical compounds, through a process called cracking. The molecules of these fractions break down and recombine as gasoline or other fractions. The cracking process produces ammonia, explosives, fertilizers, synthetic rubber, and many other compounds.

What One Barrel of Crude Oil Makes

One barrel contains 42 gallons of crude oil. The following figures are based on the 1995 average yields for U.S. refineries, as reported by the American Petroleum Institute. The total of refined products adds up to 44.2 gallons, for what is known as a "processing gain" of 2.2 gallons from every barrel of crude oil. This processing gain appears to come from liquids that are extracted from solids during the heating process.

Product	Gallons per Barrel of Crude Oil
gasoline	19.5
distillate fuel oil *	9.2
kerosene-type jet fuel	4.1
residual fuel oil **	2.3
liquefied refinery gases	1.9
still gas	1.9
coke	1.8
asphalt and road oil	1.3
petrochemical feedstocks	1.2
lubricants	0.5
kerosene	0.2
other	0.3

* includes both home heating oil and diesel fuel
** heavy oils used as fuels in industry, marine transportation, and for electric power generation

Petroleum uses

Refined petroleum compounds are used in approximately three thousand different products. Many kinds of plastics, synthetic fibers for clothing, soaps, insecticides, rubber tires, paints, and vinyl phonograph records all use petroleum products. Thus, many petroleum products are vital to both our economy and to our lifestyle.

Oil consumption in the world

Oil is clearly useful, and the world uses a lot of it. The more humans use petroleum products in their day-to-day lives, the more oil is drilled, transported, and refined around the globe. As more oil is used, more of it seeps, leaks, and spills into our environment. Even nations that have most of their own oil supplies still transport oil within their borders, and many of them export oil to other nations. At every stage of the petroleum process—drilling, transporting, refining, or normal usage—oil spills can, and do, occur.

2

Oil in the Environment

RUNNING ACROSS A parking lot on a rainy day, you might notice a rainbow at your feet. The iridescent colors swirl across the glistening pavement. Even though it's beautiful, it's the result of an oil spill.

Oil spills occur constantly, but people usually don't pay attention to them. Often, only large drilling or transporting accidents capture the attention of the media and the general public. Yet within the United States, small oil spills are the norm. According to an information sheet published by the National Oceanic and Atmospheric Administration (NOAA), an agency of the U.S. government that deals with many environmental issues, including oil spills, "Most oil spills in U.S. waters involve fewer than 100 gallons. Of the 170,000 U.S. oil spills from 1973 through 1993, over 90% involved fewer than 100 gallons. Less than 1% were spills of 100,000 gallons or more."[4]

These same figures hold true around the world. Most oil enters our environment in quantities much smaller than one hundred thousand gallons. From drilling to transporting, refining, human usage, and disposal—many opportunities exist for oil to enter our environment. Day by day, spill by spill, a large quantity of oil pollutes the environment.

Natural leaks

Humans aren't the only cause of oil in our environment. Much oil enters the environment naturally. Nature spills

Since oil is largely imported to the United States from other sources, the risks of accidents (such as oil spills) occurring in its transportation grow larger every year.

nearly 58.8 million gallons of oil into the environment every year. About one-fourth of this amount escapes from oil-bearing rock formations as they erode on land. The other three-fourths comes from cracks in the ocean's floor. More than 44 million gallons of oil seep into the ocean every year—naturally.

Along the West Coast of North America, earthquakes have fractured hundreds of faults in the sedimentary rocks of the continental shelf. Nearly one hundred natural oil seeps exist in the offshore waters near Santa Barbara, California. Every year more than 8 million gallons of oil are released from a particularly large natural seep off the coast near Santa Barbara. All told, scientists have located more than two hundred natural oil seeps off the West Coast.

Daily human leaks

Oil also ends up in our environment through a variety of daily human activities. Every day humans spill engine oil, gasoline, paint, and other petroleum products. The amount of oil spilled into the environment each time is usually minimal, often less than a gallon, but every small human spill adds up. According to research done by the U.S. Environmental Protection Agency (EPA), people who change

Oil is in constant use by various world populations, and millions of gallons each year are disposed of unsafely.

their own car oil pour approximately 175 million gallons of used oil down drains or into landfills. Much of this oil will make its way into rivers and streams, eventually reaching the ocean.

Drilling blowouts

Another way oil leaks into our environment is at the actual source of our petroleum—oil wells themselves, and the drilling process. In the early days of oil drilling, wells often spewed thousands of barrels of oil onto the ground before they were capped and contained. Even after they were capped, they could still blow out. Today fewer drilling accidents that result in spills occur than in those early days of the oil industry.

Any well can blow out when high-pressure gas and oil are encountered in the drilling process. When that happens, the oil and gas rush to the surface and flow unchecked. During a blowout, oil may spray hundreds of feet in the air, coating everything in the vicinity with sticky, smelly oil. On March 10, 1910, for example, a well called the Lake View No. 1, also known as the Lake View Gusher, blew out in Kern County, California. The well "flowed uncontrolled at rates of as much as 68,000 (barrels per day) until September, 1911, when it ceased flowing, apparently because of a bottomhole caving."[5]

Accidents like the Lake View Gusher are infrequent in modern times. As oil has become more valuable, oil companies have developed methods to recover as much oil as they can. Nowadays, drilling rigs are built with safety shutoff valves that help to control the flow of oil. Oil that is spilled on the ground during drilling can often be vacuumed up, filtered, and used. Despite precautions, however, drilling and well accidents still occur.

For example, the Ixtoc-I oil well blew out in 1979. The Ixtoc-I was an exploratory offshore well in the Gulf of Mexico. By the time the well was finally capped off, 290 days later, between 475,000 and 600,000 metric tons (or at

The "Top Five" Spills

Year	Location	Cause	Tons*
1979	Ixtoc-I oil well, Gulf of Mexico	blowout	475,000–600,000
1979	*Atlantic Express* and *Aegean Captain,* Caribbean	collision	300,000
1983	*Castillo de Beliver,* near Cape Town, South Africa	fire	250,000
1978	*Amoco Cadiz,* off the coast of northwest France	grounding	223,000
1967	*Torrey Canyon,* off the coast of southwest England	grounding	119,000

* One metric ton x 294 = U.S. gallons of oil
Laurence Pringle, *Oil Spills.* New York: Morrow Junior Books, 1993.

least 139.7 million gallons) of oil had spewed into the ocean. To date, the Ixtoc-I is the worst accidental oil spill ever. Because the Ixtoc-I was a deepwater well far from shore, most of the damage done was not visible. Oil did wash up on Texas beaches, but most of the damage was to deepwater marine life: fish, sea turtles, dolphins, and other creatures. By comparison, the *Exxon Valdez* only spilled about 37,000 metric tons of oil, but it was in shallower water and was closer to shore.

Deliberate spills

In recent years the world has seen a new kind of oil well "accident": deliberate sabotage. For example, during the eight-year war between Iran and Iraq in the 1980s, Iraq deliberately attacked Iran's oil-production facilities. By firing rockets at Iran's offshore oil wells, Iraq caused hundreds of

Kuwaiti oil wells sabotaged by Iraq burn out of control during the Persian Gulf War.

tons of oil to daily gush into the Persian Gulf. Most of these gushers were not controlled or capped for several months. Iraq used similar tactics against Kuwait during the Persian Gulf War, releasing more than 252 million gallons of oil into the Persian Gulf.

Transportation leaks

Transporting oil offers yet another opportunity for oil to spill into our environment. With the United States alone consuming more than 18.2 million barrels of oil a day, huge amounts of oil are constantly being transported around the globe. Pipelines are one way that oil is transported, and occasionally pipes break and oil spills into the environment. According to the American Petroleum Institute, in 1996 there were 169,435 miles of pipeline in the United States, offering many opportunities for breaks and resulting spills. The Trans-Alaska Pipeline stretches more than eight hundred miles across Alaska, running from the North Slope oil fields to the port town of Valdez on Prince William Sound. Every day as much as 2 million gallons of oil run through the pipeline. There have been some small spills involving less than 100 gallons along the Alaskan pipeline, but to date, no large spills have occurred. Other countries have not been so lucky, though.

In South America, Colombia also has a pipeline, which runs 430 miles from the Cano Limon oil field in Arauca to the Caribbean coast. The Colombian pipeline has been the subject of many deliberate attacks by rebel guerrillas; some of these attacks have resulted in spills. The pipeline was bombed sixty-one times in the first ten months of 1998. In June 1998 alone, six bombs were detonated along the pipeline, causing four ruptures. According to the *Oil Spill Intelligence Report,* a trade publication of the petroleum industry, "Nearly 94% of Colombia's pipeline spills can be attributed to attacks by various guerrilla factions of the National Liberation Army."[6]

For more than a decade, Colombian rebels have targeted the pipeline and other oil-production facilities. Almost every attack on the pipeline spills at least several hundreds of gallons of oil. As a result, enormous damage has been done to the environment along the pipeline. Attacks frequently cause the oil to catch fire, creating more damage. In October 1998 rebels bombed the Ocensa pipeline in northern Colombia. Besides spilling thousands of gallons of oil, the resulting fire killed more than forty-five people.

Mathematics and Oil Spills

When an oil spill occurs, many different units of measurement may be used to describe the size of the spill. This can be very confusing for the public since media reports often use the largest figure to describe the oil spill.

barrels × 35	=	imperial gallons
barrels × 42	=	U.S. gallons
cubic meters × 264.2	=	U.S. gallons
cubic feet × 7.481	=	U.S. gallons
metric tons × 294	=	U.S. gallons

Australian Maritime Safety Authority, Marine Environment Protection, "Mathematics and Oil Spills," available from http://www.amsa.gov.au/me/edu/maths.htm.

Pipeline spills on land are not usually the result of sabotage; however, they definitely add up. For example, between 1968 and 1996, overland pipelines in the United States spilled more than 181 million gallons of oil—more than seventeen times the volume of the 1989 *Exxon Valdez* tanker spill. On the average, ninety-nine pipeline spills happen each year in the United States alone.

Daily spills in the oil industry

Spills occur on a daily basis within the oil industry. Some are accidental, some are deliberate, such as when cruise ships illegally dump used oil overboard or tankers illegally pump out seawater used to clean their holds. Only about one-twentieth of all oil spills that occur during transportation are truly accidental. Most of the spills occur through human error during routine transportation procedures. Author Wesley Marx indicates that "two-thirds of the spills that occur in harbors take place during petroleum transfer operations, such as coupling and uncoupling hoses."[7]

Immense supertankers can quickly transport millions of gallons of oil but are hindered by their size. It takes many miles for them to turn around or stop.

In addition to pipelines, tankers also move large amounts of oil. In order to move large quantities of oil, supertankers, often the length of three football fields, have been developed. These tankers have the advantage of moving millions of gallons of oil at a time, but their size is also a disadvantage. Because they are so large, it takes them several minutes and several miles to turn or come to a stop.

On any given day, more than fifty tankers arrive in ports in the United States. Some tankers come directly to the dock, but many of these ports don't have docking facilities for ships that large. The tankers then anchor offshore and smaller oil tankers come out to meet them, remove their cargo—oil—and take it to the dock. During this process of removing oil from a tanker, called lightering, oil can be spilled.

The "Small" Spills of 1990

Gallons	Where	Why
2,000,000	Suez Canal, Egypt	Tanker ran aground
294,000	Toledo, Colombia	Storage tank bombed by guerrillas
248,700	Pacific Ocean, Japan	Tanker sank in typhoon
237,000	Wakasa Bay, Japan	Tanker split in two
164,000	Hudson River, New York	Tanker hit reef
100,000	Perth Amboy, New Jersey	Storage tank collapse
75,000	Freeport, Pennsylvania	Pipeline rupture
75,000	Minnesota River, Minnesota	Pipeline malfunction
47,300	Manchester, Washington	Storage tank leak
21,000	Lake Charles, Louisiana	Pipeline weld failure
16,800	Oahu, Hawaii	Collision with buoy
12,600	Corpus Christi, Texas	Leak in tanker
10,000	Tenerife, Canary Islands	Tanker struck dock
10,000	Burrard Inlet, Canada	Collision with barge

Adapted from Laurence Pringle, *Oil Spills.* New York: Morrow Junior Books, 1993.

When oil spills during lightering, it's usually because of human or mechanical error. Although the amount of oil spilled is usually small, enough oil is spilled during lightering that the U.S. Coast Guard has requested the National Academy of Sciences to study the problem and make recommendations. The academy is conducting "studies on oil spill risks from lightering operations off the coast of the United States and on the efficacy of automatic fuel shut-off equipment to prevent oil discharges during transfer operations." The study will also "evaluate current lightering practices and trends and analyze the associated risks."[8]

When all the oil has been offloaded, either to a smaller tanker or directly at the dock, approximately 1 percent of a tanker's cargo still remains in the hold. To clean their holds and remove the remaining oil, tankers pump seawater into the hold, then pump it into special holding tanks. In these tanks, the oil separates from the seawater and rises to the top, where it is skimmed off and added to the rest of the oil shipment or to the next cargo.

Although this process works well for retrieving most of the oil, some oil remains behind in the seawater that was used for cleaning. What happens to this produced water, as it is called, often adds yet more oil to the environment.

International agreements regulate the disposal of produced water and state that there can be no more than 15 parts oil per 1 million parts seawater in any produced water that is discharged back into the environment. But with so many tankers coming and going on a daily basis—all of them cleaning their holds—it is a difficult regulation to enforce. The National Academy of Sciences estimates that up to 3,000 metric tons (882,000 gallons) of oil are dumped into the environment every year through produced water.

It's also difficult to remove all the oil from a cargo hold. Tar fractions of petroleum often stay behind when a tank is cleaned. Tankers often wash their holds again, when they're far out at sea, discharging these sticky tar residues into the open ocean. This practice is illegal, but as tanker traffic increases, so do the reports of sticky tar masses washing up on previously unpolluted beaches, like those on Bermuda and on the east coast of Africa. Officials know that illegal tanker discharge is the cause of these problems.

Thanks to recent technological developments, oil itself can aid investigators in helping to determine the origin of the spilled oil. The hydrocarbons in oil have their own unique composition and can be identified. Oil pumped from a Saudi Arabian oil field is different from oil pumped from an Alaskan oil field. Although oil is basically the same for refining and consumer purposes, thanks to hydrocarbons each oil field is different and identifiable—almost the way a human fingerprint can identify a person.

When scientists examine illegally pumped oil, they can tell which oil field it came from; consequently, due to shipping records, they can often identify which vessel was carrying the oil and fine the company that owns it.

Dramatic spills

Although large tankers account for only a small percentage of oil spilled into our environment, it is these large accidents that capture the public's attention. The sight of millions of gallons of oil spewing into the ocean, usually due to human error, is certainly dramatic. Little spills may add up, but large spills are impossible to ignore.

Since 1990 tanker accidents have actually declined, in part because the penalties imposed by the U.S. government for environmental damage have become much heavier. As a result of the Oil Pollution Act of 1990, an oil spill caused

Penalties imposed by the U.S. government have limited the number of recent oil spills, but they can still happen.

by carelessness can now cost a petroleum company hundreds of millions of dollars in fines and cleanup costs.

But tanker accidents can, and do, still happen. Tankers can break apart in rough seas, run aground in shallow water, or crash into a reef. In 1978, after the steering mechanism on the *Amoco Cadiz* failed, the ship ran aground in shallow shoals off the coast of France and broke in half. More than 65 million gallons of oil contaminated two hundred miles of French coastline.

Because large tankers take so long to turn, slow down, or stop, they can even collide with other ships. In 1979 the *Atlantic Express* and the *Aegean Captain,* two large tankers, collided off the coast of Trinidad, spilling more than 88 million gallons of oil into the pristine waters of the Caribbean.

Likewise, in 1989 the *Exxon Valdez* ran into a reef in Prince William Sound, Alaska, due to human error and inattention. On a perfectly clear night, and in a totally calm sea, more than 10 million gallons of crude oil spilled into Prince William Sound.

Silent, deadly spills

Sometimes oil seeps into the environment without anyone noticing, such as when an underground storage facility develops a leak. Although not as dramatic as a tanker crash or a gushing well on land, this type of spill can be much more harmful because it can go undetected for years.

In Brooklyn, New York, at a storage facility owned by Mobil Oil Corporation, more than 17 million gallons of petroleum—everything from crude oil to kerosene and gasoline—leaked into the ground from underground pipes and storage tanks over a period of nearly forty years. Officials estimate that the oil leaked continuously from the 1940s through the 1970s. Because the leaks were going directly into the soil, no one saw the leakage or environmental damage until it was too late. The petroleum contaminated residential neighborhoods, creeks, and groundwater before anyone noticed.

Underground storage tanks at this Mobil facility in Brooklyn leaked more than 17 million gallons of petroleum into the ground.

In 1971, at Searsport, Maine, five thousand gallons of jet fuel and heating oil leaked from underground storage tanks into Long Cove. The problem was only noticed when a population of soft-shelled clams, which had always thrived in the cove, died off.

Underground petroleum leaks are extremely difficult to clean up, particularly when they involve refined oil. The effects of underground oil pollution can linger in the environment for years. Mobil Oil Corporation, the company primarily responsible for the Brooklyn spill, has spent years and millions of dollars trying to remove the oil from the ground. Ten years after the petroleum had leaked into Long Cove, the cove still hadn't recovered from the environmental damage.

All these spills, large and small, put millions of gallons of oil into our environment every year. What happens when all this oil spills?

3
Oil Spills

WHENEVER OIL SPILLS into our environment, damage occurs. But the amount of damage done depends on many factors. Salt- or freshwater, water temperature, shoreline contour, nature of the oil spilled, air temperature, wind and wave action, season of the year, amount of sunlight, all of these variables affect the severity of the spill. The list of factors affecting damage is endless, which is why oil spills pose so many challenges. No hard-and-fast set of rules exist for what happens during an oil spill; no two oil spills are alike.

Type of oil

The most important factor in any oil spill is what type of oil has been spilled. Crude oil acts differently than jet fuel, kerosene, or gasoline, for example.

In the 1970s two oil spills occurred in Maine within one year of each other. Each one involved a different type of oil. In one spill, five thousand gallons of jet fuel leached through the ground into Long Cove. In the other, a tanker scraped a reef and spilled forty thousand gallons of fuel oil into Casco Bay.

To a casual observer, the forty-thousand-gallon spill would appear to be far worse than the five-thousand-gallon spill. But ten years after the spills, it was almost impossible to find any signs of the oil spill in Casco Bay or any signs of permanent ecological damage there. In Long Cove, though, massive ecological damage was still evident. The major difference between the two spills was not

Reprinted by permission of Greenpeace.

the amount of oil spilled but the type of oil spilled and the environment in which it spilled. Five thousand gallons of jet fuel leaching through the ground into a small cove was far more toxic than forty thousand gallons of fuel oil spilled in open water.

Although the petroleum industry classifies oil according to its geographical location, for example, "Alaska North Slope Crude," the U.S. Environmental Protection Agency (EPA) has developed four distinct classifications for petroleum products. The classifications were developed to help cleanup crews during an oil spill. The toxicity and cleanup procedures are different for each classification. By knowing which category they're dealing with, cleanup crews can work more efficiently.

The first category, Class A, deals with light, volatile oils that are very liquid, have a high evaporation rate, are usually flammable, and are extremely toxic to all living things. Jet fuel, gasoline, and other quality light crudes fall into this classification.

Class B deals with nonsticky oils that have a waxy feel. Class B oils are less toxic than Class A ones, but they adhere more firmly to surfaces. However, vigorous flushing and hosing can often remove this type of oil. Medium to heavy paraffin-based oils fall into this class.

The third classification, Class C, involves heavy, sticky, and tarry oils. This class of oil often sinks, and flushing or hosing rarely removes the oil. Class C oils are not very toxic, but they're so heavy that wildlife that comes in contact with the oil can be smothered or drowned. Class C oils include medium to heavy crudes and some residual fuel oils.

Class D deals with nonfluid oils that are not very toxic but are quite heavy and difficult to clean up. When heated, Class D oils will actually melt and coat surfaces, making it nearly impossible to clean the object. Heavy crude oils, some high-paraffin oils, and some weathered oils fall into Class D.

Although the classifications help cleanup crews know what they're initially dealing with, every oil spill is unique. Even with all the research that went into determining the EPA classifications, it's still hard to know exactly how oil will behave during a spill.

Oil spills

Oil spills can occur almost anywhere, but most major oil spills occur in the ocean, when oil is being transported by tanker. The ocean is what concerns most people when it comes to an oil spill. A spill that occurs on land is easier to contain and clean up because the oil doesn't disperse as rapidly as it does in water. An oil spill in the ocean usually causes more widespread environmental damage than a spill on land.

The top few feet of the ocean are its most productive. Plankton, tiny plants and animals that are the basis of the oceanic food chain, live in this part of the ocean. All sea life depends on plankton, directly or indirectly. Fish eggs and fry (newly hatched fish), crab and lobster larvae, young turtles—most of the ocean's young, in fact—also live in the upper few feet of the ocean. This vital layer, the ocean's "skin," is precisely what an oil spill devastates.

Because these top few feet of the ocean are indeed the basis of the oceanic food chain, even a small oil spill

Reprinted by special permission of King Features Syndicate.

causes problems throughout the entire chain. Animals that aren't killed outright by the oil spill are still contaminated. As they are eaten by other creatures, those creatures also become contaminated.

How oil behaves with water

Since oil is lighter than water, it tends to float on top of seawater. Thus, even if a spill starts underwater, such as when a tanker hull breaks open or an offshore well develops problems, the oil still rises to the water's surface. When oil hits the surface of water, it immediately starts spreading.

The speed and pattern of spreading oil depends primarily on the water temperature, currents, and wind. Of course, the amount of oil spilled also affects how much it spreads. In the 1970s, a tanker spilled 8 million gallons of thick crude oil off Nantucket, an island in the Atlantic off the coast of Massachusetts. At first the spill was a thick, gooey mat only one mile square. But wind, current, and temperature conditions turned the spill into a thin oily sheen covering more than eleven thousand square miles.

Weathering: Nature helps clean up

When oil spills into water, natural processes begin to affect the oil in a series of events known as weathering.

According to the EPA, "weathering is a series of chemical and physical changes that cause spilled oil to break down and become heavier than water."[9] The severity of an oil spill may be reduced somewhat by these processes.

As soon as oil starts spreading, it gets exposed to air, and different fractions of it start to evaporate—one of the first steps in weathering. The most lightweight fractions of petroleum, such as benzene and other extremely poisonous compounds, turn into vapor and leave the water. Within a few hours of the spill, lighter petroleum fractions will already be evaporating. As spreading and evaporation continue, heavier fractions may also evaporate, though not as much or as quickly as the lighter fractions. Within two weeks, as much as half of an oil spill may have evaporated into the air. In the case of extremely light fractions, such as jet fuel, kerosene, and gasoline, the spill may evaporate completely.

As evaporation continues, the remainder of the oil spill becomes dense, heavy, and sticky. Tar balls and floating mats of dense, sticky oil may form on the surface of the water. Often, this material settles on the ocean floor or washes onto beaches.

Oil-eating bacteria

Not all life forms find oil to be toxic. According to the Marine Environment Protection department of the Australian Maritime Safety Authority, "Many species of marine micro-organisms or bacteria, fungi and yeasts feed on the compounds that make up oil."[10] This process is called biodegradation. When microorganisms consume oil, they metabolize petroleum hydrocarbons into carbon dioxide and water. Microorganisms are one of nature's best defenses against oil spills.

How fast biodegradation occurs depends on the temperature of the water and oil mixture. Biodegradation works fastest in environments where the water is warm and contains lots of oxygen. Nutrients in the water, like phosphorus and nitrogen, are also helpful. But even under the best of conditions—where the water is warm and nutrient-rich—biodegradation is a slow process, often taking years.

Poisonous leftovers

With oil evaporating and microorganisms munching away, it might seem pointless to worry about a spill or even to try to clean it up—just let evaporation, biodegradation, and the rest of the natural weathering process do the job. Unfortunately, the problems created by spilled oil are not so easily solved.

Nature does do a good job of cleaning up an oil spill, but nature doesn't dump millions of gallons of oil in one place at one time, the way a tanker accident does. Millions of gallons are too much oil for the environment to handle at one time. Natural processes are also best suited for handling crude oil, not all the toxic derivatives that are refined from it. For example, although jet fuel may eventually evaporate completely, during the process the fumes are toxic to any living thing encountering them—seals, otters, whales, birds, and humans. And although crude oil in its natural state is the least toxic of all petroleum forms, it is still toxic.

Oxidation is also part of the weathering process and "refines" spilled oil into even more toxic substances. According to the EPA, "Oxidation occurs when oil contacts the water, and oxygen combines with the oil to produce water-soluble compounds."[11] Among these compounds are alcohols, ketones, and peroxides, and they cause far more harm

Although nature can clean up small amounts of oil, it can not handle the massive consequence of an oil tanker spill.

to marine life than the original petroleum fractions. Because they're water soluble, they can't be gathered up; thus, they remain in the environment, often causing damage for years to come. Some oxidation occurs around the edges of an oil slick, no matter what type of petroleum is involved or how fast it is cleaned up.

When oil and water mix

During an oil spill, oil doesn't always stay on the surface of the water and spread. If the wind, waves, and water temperature work together in a certain way, the oil and water will actually mix together. The action of wind and waves will push the oil droplets beneath the surface of the water, spreading the oil much farther and deeper than the thick oily coating on the surface would indicate.

Although the oil may not be as visible in this type of situation, that doesn't mean it has evaporated, biodegraded, or weathered in any other way. The oil is still there, and enormous damage to the environment is still being done.

Where Does the Oil Go?

In 1978 the *Amoco Cadiz* ran aground off the coast of Brittany, France, spilling more than 223,000 tons of crude oil. Because the accident occurred near several oceanographic laboratories, the *Amoco Cadiz* remains one of the best, and most extensively, studied spills of all time. In studying the disaster, scientists discovered that:

— 30 percent of the oil evaporated;
— 14 percent dispersed directly into the water;
— 8 percent was deposited, through sedimentation, on the seafloor;
— 4 percent was consumed through biodegradation; and
— 28 percent washed up into the intertidal zone.

High waves were present at the time of the stranding and whipped a large amount of the oil into "chocolate mousse." More than two hundred miles of France's coastline was fouled by the thick, gooey oil.

Emulsions

When winds, waves, and currents are strong enough to whip water droplets into oil, such as when a tanker breaks up in a storm or heavy surf, a process called emulsification occurs. Two kinds of emulsions exist: water-in-oil and oil-in-water.

Water-in-oil emulsions are easily visible. As more and more water becomes trapped in the oil, a frothy pudding-like mixture called chocolate mousse forms.

Oil-in-water emulsions are difficult to see, almost invisible to a casual observer. Consequently, they can be even more harmful than chocolate mousse. According to the EPA, the U.S. agency responsible for monitoring and cleaning up oil spills, "Oil and water emulsions cause oil to sink and disappear from the surface, which give the false impression that it is gone and the threat to the environment has ended."[12]

Emulsification slows the weathering process and makes cleanup difficult. Chocolate mousse may linger in the environment for months, even years. And no one really knows how long oil-in-water emulsions remain since they usually disappear from sight—yet continue to cause damage.

Sedimentation

Even when evaporation, biodegradation, and weathering have reduced an oil spill to sticky tar balls and mats, another problem still remains: sedimentation. Water contains sediment—bits of clay, dirt, and sand. As these tiny bits of matter mix with oil, the oil becomes heavy and sinks. Sedimentation can happen fairly quickly to heavy oils without any weathering or weeks after a spill, when the lighter fractions have evaporated.

Sedimentation keeps oil in the environment for a long time. Then, during storms or even with normal wave action, the oil-laden clay and sand deposits wash up on beaches.

Evaporation, biodegradation, and weathering all play a part in cleaning up an oil spill. But more than anything, humans play the greatest part of any cleanup operation.

4
Cleaning Up the Environment

PEOPLE VIEW OIL spills with horror, especially when gooey oil comes ashore and fouls pristine beaches or runs into streams and decimates entire fish populations. The public wants something done immediately, and it usually demands that the mess be cleaned up and the environment restored.

Cleaning up after a large oil spill is never simple or easy, and frequently, it's not immediate. Even the best efforts often can't entirely fix the problem—the scope of most oil spills is just too large. Sometimes, the remedies employed to clean up a spill actually make the problem worse.

Because oil is essentially liquid, it travels like liquid—it can go almost anywhere. When oil is spilled in water, it can travel even more extensively because it can go wherever the water goes. Regardless of where a spill occurs, the oil may end up on sandy beaches, rocky coastlines, mingled into the reeds and sediments of saltwater marshes, or coating the shoreline of freshwater estuaries. Every place that oil settles poses a different cleanup problem.

When oil is spilling, the first thing cleanup crews want to do is shut off the source of the oil. Sometimes, such as when the spill involves a pipeline, the spill can be stopped by simply closing a valve. But it's not so easy to stop the spill if it's coming from a damaged tanker. Sometimes a leaking hull can be patched; usually, however, the best that can be done is for the owner of the cargo to try and save the remaining oil by lightering it onto smaller vessels. Once oil spills, cleanup crews are first concerned with containing the spill as much as possible.

Booms

If oil is spreading slowly, seas are calm, and cleanup crews act quickly, some of the spilled oil can be recovered, sometimes before it does much damage. Tubes or poles, called booms, are linked together and placed around the edges of the spill, much like a floating fence. Booms help to keep the oil in one place.

Booms, or devices that "fence in" oil, can be used in calm water to contain an oil spill.

All booms share four features: 1) an above-water "freeboard," which helps prevent small waves from splashing oil over the top; 2) a below-water "skirt" to contain oil below the boom as much as possible; 3) a flotation device to keep the boom on the surface; and 4) an anchoring device, which keeps the booms and the oil they surround in one place.

Although some booms are designed especially for rough water, they work best in calm water where the surface is smooth and there's little current. Booms don't work at all when currents, wind, and waves are present—the booms bounce around, and waves wash the oil over and under them. According to the EPA,

> While most booms perform well in gentle seas with smooth, long waves, rough and choppy water is likely to contribute to boom failure. . . . Generally, booms will not operate properly when waves are higher than one meter or currents are moving faster than one knot per hour.[13]

Booms require constant tending and maintenance, and they need to be moved as the tides move the oil slick.

Equipment is not always immediately available when an oil spill occurs. The EPA stresses that containment is critical; people on the scene of a spill should use anything they can lay their hands on to keep the oil from spreading:

> Improvised booms are made from such common materials as wood, plastic pipe, inflated fire hoses, automobile tires, and empty oil drums. They can be as simple as a board placed across the surface of a slow moving stream, or a berm built by bulldozers pushing a wall of sand out from the beach to divert oil from a sensitive section of shoreline.[14]

In 1989 booms helped to contain thousands of gallons of the oil spilling from the *Exxon Valdez*. The water in Prince William Sound was exceptionally calm at the time of the spill, and workers were able to surround the stricken tanker with booms while they lightered off the remaining oil. Even so, as the full impact of the *Exxon Valdez* disaster demonstrates, booms are only part of the solution.

Booms and suction devices can help remove the main portion of oil in natural areas. Here, workers attempt to clean oil spilled from a Russian tanker near the shores of Mikuni, Japan.

Skimmers

If the oil can be contained, the next task workers face is to get it out of the water. A skimmer is a device used to retrieve oil from the water's surface. Working somewhat like a household vacuum cleaner, skimmers can be used from shore or from a boat.

Boats equipped with mechanical skimmers can recover some of the oil from the boom-enclosed area and get it back into storage tanks. Like booms, skimmers work best in calm seas. In rough or choppy seas, skimmers suck up more water than oil. Skimmers also need to be constantly tended and cleaned since, in addition to oil, they also suck up any debris or ice that's in the water.

Sorbents

Sorbents, which are materials that soak up liquids, are another way of gathering up spilled oil. Cleanup workers spread them over the surface of an oil spill so that they soak up the oil. The sorbents are later gathered up, and the oil is squeezed out.

Sorbents gather oil in two different ways, depending on the material from which they are made. According to the EPA, "**Ab**sorbents allow oil to penetrate into pore spaces in the material they are made of, while **ad**sorbents attract oil to their surfaces but do not allow it to penetrate into the material."[15] For example, a paper towel would be an absorbent, and a wooden board would be an adsorbent. Adsorbents are usually easier to clean and reuse, but they can be more expensive and cumbersome to use. Some absorbents are harder to clean up, but most types are inexpensive and easy to use.

Many different materials may serve as sorbents. How much oil they will collect depends on whether the sorbent is made of natural organic, natural inorganic, or synthetic substances.

Examples of natural organic sorbents are straw, hay, peat moss, sawdust, feathers, and ground-up corncobs. The advantage of natural organic sorbents is that they are usually inexpensive and are readily available, and they can soak up between three and fifteen times their weight in oil. The disadvantage of natural organic sorbents is that they tend to float loose (like sawdust) and sink when they become soaked with oil. Even so, natural organic sorbents are used more frequently than other materials to clean up oil spills. To combat their disadvantages, cleanup workers wrap natural organic sorbents in mesh and then anchor them to flotation devices. It's not unusual to see sawdust or straw wrapped in mesh to contain it and tethered to an empty floating oil drum to keep it from sinking.

Natural inorganic sorbents include materials like sand, clay, volcanic ash, vermiculite, glass wool, and perlite. These materials can absorb between four and twenty times their weight in oil. Like natural organic sorbents, they are also readily available and inexpensive, but they tend to be difficult to contain and to sink once they are saturated with oil.

Synthetic sorbents can absorb up to seventy times their weight in oil, and many kinds can also be cleaned and

An oil spill worker uses sorbents to absorb the toxic oil. These spongelike materials can hold many times their weight in oil.

reused. Synthetic sorbents include such human-made materials as polyurethane, polyethelene, and nylon fibers. Although synthetic sorbents are manufactured in many different forms, the most common is sheet form. Synthetics are lightweight, easy to use, and often are reusable; however, synthetics are usually more expensive than other sorbents.

Gelling agents

Gelling agents, sometimes called solidifiers, are chemicals that react with oil, turning it into a rubberlike substance that can be gathered up by skimmers or nets. In the case of a small spill, gelling agents can be applied by hand and left to mix with the oil on their own. With larger spills, the chemicals need to be applied to the slick using high-pressure hoses. The force of the chemicals coming out of the hose helps mix them into the oil.

Gelling agents work well with small spills in moderately rough seas. The wave action mixes the oil and gelling agent, eliminating the need for high-pressure hoses.

The main disadvantage of gelling agents is that a large amount must be applied to the spill in order for them to work, sometimes as much as three times the amount of spilled oil. This makes gelling agents impractical for large spills. If a spill is 1 million gallons, for example, it's not practical to store, move, and apply 3 million gallons of gelling agent.

Burning

Burning is another tool in cleaning up an oil spill. In burning, the oil is ignited on the surface of the water. Although no oil is recovered with this method, no oil is left to destroy beaches and aquatic life.

Burning works best when a spill is fresh. In freshly spilled oil, the lighter, highly flammable fractions, such as gasoline and kerosene, are still present and allow the oil to ignite. Once the lighter fractions evaporate, the spill is harder to burn; in such cases, gasoline or kerosene has to be added to the slick to make it ignite.

Burning can only be done in open waters where there is no danger to other ships, drilling rigs, or the shoreline. A fire boom is placed around the spill to contain it during the burning. In addition to the dangers posed by the fire, often substantial air pollution is a problem.

Helping biodegradation

Biological technologies are also used on oil spills. Some biological agents—certain chemicals and bacteria—will actually aid and increase the natural process of biodegradation.

Natural biodegradation of a large oil spill can take years, but when it comes to ecologically sensitive shorelines and wetlands, this isn't fast enough. The addition of biological agents can speed up the biodegradation process and help minimize damage to aquatic habitats.

Two different biological techniques are currently being used in the United States to help clean up oil spills: fertilization and seeding. Both of these methods speed natural biological processes.

In fertilization, nutrients that also occur naturally, such as phosphorus and nitrogen, are added to the water. Adding nutrient chemicals to the water stimulates the growth of existing microorganisms, some of which feed on petroleum products. Fertilization thus increases the natural rate of biodegradation.

Seeding is the process by which more microorganisms are added to the environment. Native populations of microorganisms may already exist in the area of an oil spill, but their numbers are usually not large enough to handle an oil spill of hundreds of thousands of gallons. Cleanup crews add foreign bacteria to the water, which increases the population and thus speeds biodegradation.

By adding more oil-eating bacteria to the water, and then giving them nutrients to increase their growth, workers can help speed the "natural" cleanup of an oil spill. Because this is new technology and involves chemicals and bacteria, workers must monitor it carefully. Seeding and fertilization have produced some promising results in sensitive areas like shorelines, marshes, and wetlands.

Dispersants

Sometimes, as in the case of extremely light oils, a spill spreads too thinly to be contained. In those cases, instead of trying to retrieve the oil, cleanup workers do just the opposite: They try to disperse it.

Dispersants are chemicals that react with liquid substances, such as oil, and break them into tiny droplets, similar to the way weathering thins out and breaks up a spill. The smaller broken-down droplets then mix with the water, and winds, waves, and currents break them down even further. Thus, dispersants help clear oil from the surface of the water, making it less likely that the oil slick will reach the shoreline. Often a dispersant is nothing more than a strong detergent, such as the kind used to wash laundry and dishes.

In the United States dispersants are controversial and are rarely used. Many U.S. scientists disagree about the effectiveness of dispersants; moreover, they believe that dispersants are toxic and end up damaging the environment more than the spilled oil does. As the EPA points out, though, "Dispersants used today are much less toxic than those used in the past."[16] However, dispersants are still considered to be so toxic that federal authority must be obtained in each case before they may be used.

In 1967 one of the largest tankers in the world, the *Torrey Canyon,* stranded and then broke up off the coast of Cornwall in the United Kingdom. The *Torrey Canyon* spill was the first major case in which detergent/dispersants were used—with horrible results. A fleet of vessels sprayed detergent onto the oil, then churned it up with their propellers. Within the first weekend of the disaster, all the available stocks of detergents in the British Isles had been used up, and the oil slick had turned into chocolate mousse. In the end, countless animals and birds were killed, and the oil remained. Dispersants have been used on spills since the *Torrey Canyon* disaster with similar results.

To overcome the objections of scientists and avoid known problems, new dispersants are being developed.

The oil tanker Torrey Canyon *broke in two off the coast of England in 1967, releasing millions of gallons of oil into the water.*

These new technologies are being tested in laboratories and in actual spill situations. Scientists and oil companies are gathering data in an effort to make dispersants perform better.

When oil hits a shoreline

The major concern of most cleanup workers is to keep an oil spill from reaching shore since shoreline environments are more sensitive and cleanup is more difficult. Unfortunately, despite efforts to contain and clean up an oil spill at sea, oil usually washes onto beaches. When this happens, even though cleanup operations with booms and skimmers continue in open water, other measures are needed.

According to the EPA, "Mechanical containment or recovery (such as booms and skimmers) is the primary line of defense against oil spills in the U.S." However, "physical methods, such as wiping with sorbent materials, pressure washing, and raking and bulldozing . . . are used to clean up shorelines."[17]

The type of coastline an oil spill hits helps determine cleanup procedures. Sandy beaches require different approaches than rocky coastlines do; marshes require still other treatments.

Sand and gravel beaches

When oil fouls a sand or loose-gravel beach, the beach itself acts like a sorbent, trapping the oil in the sand. Although the oil may appear to be gone, it has settled below the surface layers of sand or gravel.

Many of the beaches in Alaska that were affected by the *Exxon Valdez* oil spill were loose gravel. During the cleanup of the *Exxon Valdez* disaster, workers used high-pressure hot water hoses to clean some of these beaches. Although the beaches appeared to clean up beautifully, all the pressure cleaning did was push the oil deeper into the beach. Workers discovered a thick layer of oil only eight inches down.

Sometimes sorbents can be used to clean up a sand or gravel beach, but they have to be applied immediately or the beach itself will become the sorbent. In most cases, the only cleanup method that works is to bulldoze the sand and gravel and remove it.

Rocky shorelines

Rocky beaches pose a different set of cleanup problems. As the oil comes ashore, each individual rock is coated as well as anything living among the rocks—mussels and other intertidal species, such as crabs, oysters, and clams. The oil also collects in pools and natural troughs in the rocks, creating scattered puddles of oil.

The only way to get oil off of a rocky beach is to individually wipe each rock with sorbent material. Likewise, oil collected in pools and natural troughs can sometimes be suctioned up. Dispersants have also been used on rocky beaches, but with limited success. High-pressure water hoses were used on large rocky beaches during the *Exxon Valdez* cleanup, but with the same ultimate results as gravelly beaches—the oil didn't disappear, it just moved deeper under the surface of the beach.

High-pressure washing also appears to have devastating effects on the intertidal species that live on and among the rocks. Subsequent long-term studies have shown that the recovery of oiled intertidal species is slower on beaches that undergo high-pressure washing. Cleanup workers have concluded that individual wiping is the only way to truly remove oil from a rocky beach.

Salt marshes and freshwater estuaries

Marshes and estuaries present some of the greatest cleanup problems during an oil spill. Marshes usually con-

Workers clean rocks by hand on a beach in Prince William Sound.

tain thick bottom sediment and abundant plant life. As the oil slick comes ashore, it coats all the plants in the marsh and settles into the bottom sediment. It's difficult to remove oil from marsh sediment and almost impossible to clean off plants—they just die.

Marshes act as nurseries for many species of aquatic life. Everything from clams to turtles spends some part of its infancy in a marsh. Although often invisible to the naked eye, these juvenile species are an important part of the ocean's food chain. An oil spill that reaches a marsh causes problems in the food chain that will be present for years to come.

In marshes, hand cleaning with sorbents can help, but biological techniques have proved to be the most effective in helping marshes recover from the effects of an oil spill. Workers seed and fertilize the marsh with microorganisms that eat the oil. Even so, marshes often remain so polluted that other measures are called for if plants and wildlife are to recover.

In 1994, for example, five years after the *Exxon Valdez* oil spill in Alaska, thirty-eight tons of oiled sediment were removed from beneath a dozen oiled mussel beds in Prince William Sound. The toxicity level of the sediment was high, and the mussels were not recovering. Cleanup workers replaced the oiled gravel and sediment with clean sediment, which reduced the oil level at those sites by 95 percent. The sites are being monitored to see how they recover, but officials overseeing the cleanup felt that removal of the oiled sediment was the only way to help the intertidal wildlife recover.

Cleaning the cleanup materials

Once the spilled oil has been removed from the environment, there's still a lot of cleanup to be done. Anything that has oil on it needs to be cleaned, and this includes all the materials and equipment that were used to clean up the spill. Booms, skimmers, suction hoses, sorbent materials, and the protective clothing worn by workers—all have to be cleaned. Frequently, the sorbent materials and protective

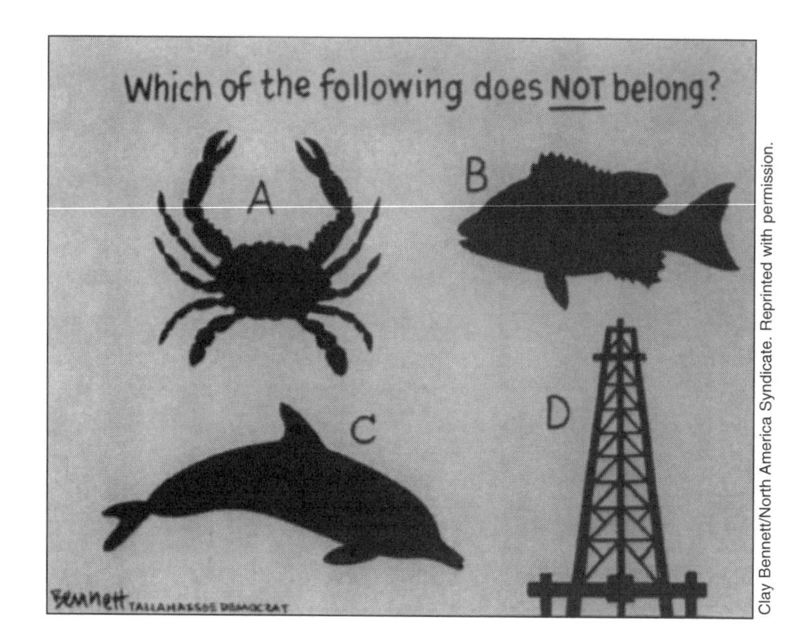

Clay Bennett/North America Syndicate. Reprinted with permission.

clothing have to be disposed of because they can't be cleaned.

Cleaning the cleanup materials poses its own set of problems. Any washing of materials creates produced water, or water that contains oil. This water needs to be either disposed of or recycled. Sorbent materials also need to be disposed of in a way that doesn't pollute the environment. Special waste sites need to be created since oil-covered materials can't just be deposited in the city dump.

The cleanup materials left over after an oil spill illustrate the main problem of such spills: Spilled oil is almost impossible to get rid of, and some of it will always remain in the environment.

Cleaning up the environment after an oil spill is only one part of the problem. Cleaning up all the wildlife and helping populations recover is another enormous task cleanup workers face.

5

Cleaning Up Wildlife

OIL SPILLS DAMAGE more than the ocean, beaches, and marshes. Any type of wildlife caught in an oil spill is affected in some way. Some animals simply die since there is no way to help them.

Fish

Fish are severely injured during water-based oil spills. Often, these life-forms suffer the most damage of all creatures affected by an oil spill. Whether the spill occurs in fresh- or saltwater, hydrocarbons either kill or severely injure fish. Humans are helpless to save fish that are injured in an oil spill.

Petroleum will usually kill fish outright by interfering with their ability to extract oxygen from the water. But even low concentrations of oil that may not kill them outright will harm fish enough that they eventually die. Low concentrations of oil in water may harm a fish's vision, retard its growth, and damage its ability to reproduce. As little as 5 parts oil per 1 million parts water will damage a fish's sense of smell, making it difficult for it to find food, thus causing it to starve to death. Fish eggs are often damaged or killed by oil as well.

Intertidal communities and other marine life

Intertidal communities are also hard-hit during an oil spill. The intertidal zone is the area of the beach that is under water when the tide is high and above water when the tide is low. Creatures that live in the intertidal zone—such as mussels, crabs, anemones, sea stars, and other shellfish and crustaceans—will be submerged in water (and oil) twice a day when the tide comes in. As with fish, there is nothing humans can do for intertidal communities once these creatures come in contact with a spill. Even wiping oil off crustaceans' shells or removing oil-saturated sand and gravel cannot completely undo the lethal effects of oil on an intertidal community.

Mussels, clams, oysters, and other shellfish may not immediately appear to be harmed by an oil spill. However, because shellfish obtain their food by filtering the surrounding water through their bodies, petroleum's effects still severely damage them. Since shellfish are unable to excrete the oil that filters through their bodies, petroleum toxins keep accumulating in their systems, quickly reaching lethal levels. The petroleum impairs the ability of shellfish to reproduce, leading to long-term population declines.

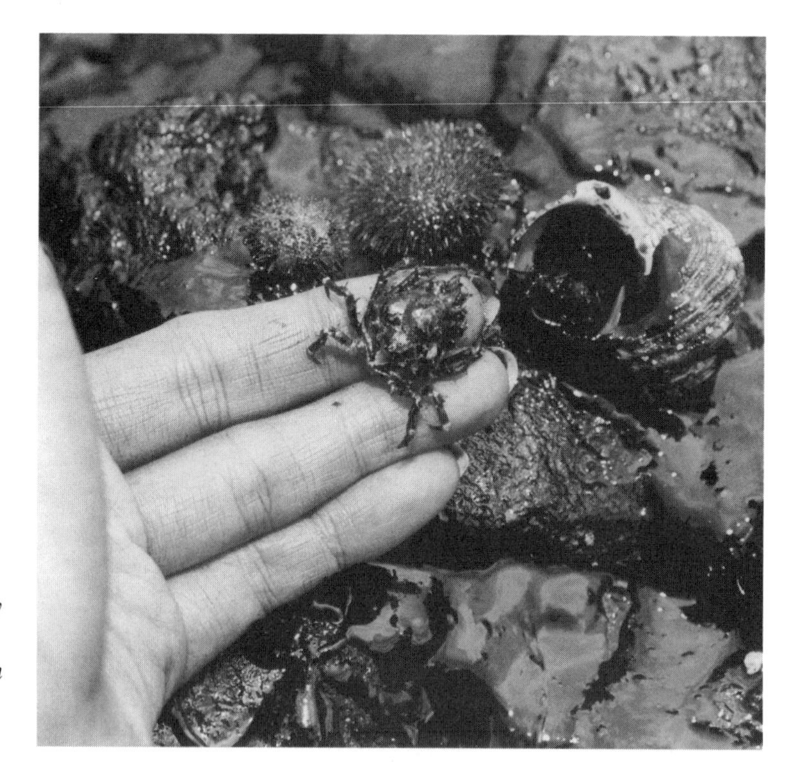

Though crustaceans may not appear to be immediately affected by oil spills, the toxins created by oil spills can lead to long-term population decline.

The sublethal effects of oil on marine life

Some effects of an oil spill are not immediately deadly but are nonetheless serious. Initially, scientists were not fully aware of the sublethal effects of an oil spill. Laboratory tests had not shown any long-term harmful effects on some animals from small doses of oil. Until recently, for example, scientists didn't think oil caused any damage to coral reefs, which comprise millions of tiny animals. But, as it turns out, coral reefs are especially susceptible to damage from oil spills.

In 1986, for example, more than 2 million gallons of crude oil spilled into the Caribbean Sea off the coast of Panama. As author Lawrence Pringle points out,

> The oil affected about fifty miles of shoreline habitats, seriously damaging the plant and animal life of mangrove forests, seagrass beds, and coral reefs. Some reefs were killed outright. Others suffered from sublethal effects, as algae and other organisms invaded their injured skeletons. Five years after the spill, just five percent of the reef was alive, and it had little chance of withstanding the eroding action of the sea.[18]

Human intervention

Although there is little humans can do for fish, intertidal communities, and coral reefs exposed to spilled oil, it is necessary for humans to intervene on behalf of animals that can be helped. According to the EPA, "When an oil spill occurs, birds and marine mammals are often injured or killed by oil that pollutes their habitat. Without human intervention, many distressed birds and animals have no chance of survival."[19]

In recent years much has been learned about both the effects of oil spills on wildlife and the care and treatment of wildlife injured in an oil spill. The EPA indicates that four conditions are essential to successfully deal with oiled wildlife:

> First the need for immediate response is essential for rescuing birds and marine mammals. Second, personnel training is needed. The rehabilitation of oiled wildlife is a complex medical and technical procedure, and volunteers must be

> properly trained. Third, a commitment must be made to reclaim oiled wildlife using proven, documented procedures, and avoiding shortcuts. Finally, open communication with other response agencies is crucial for any wildlife rescue operation to be successful.[20]

Governments and private organizations around the world have developed procedures for helping wildlife affected by oil spills. Although in the United States the federal government has overall responsibility for animal-rescue efforts following a spill, a plan is usually developed that coordinates efforts of both government and private agencies.

Once a spill has been stopped, rescue workers are confronted with two major tasks: preventing wildlife from entering the spill area and gathering up those animals that have already been exposed to the oil.

To keep wildlife from entering a spill area, workers attach helium balloons to booms, fire air cannons from boats or the beach, and anchor floating dummies close to the spill area to scare wildlife away from the site. These methods are moderately effective, particularly with birds, but they clearly don't help any wildlife that has already been oiled. Oiled animals must be captured and cleaned.

Rescuing and cleaning birds

While all wildlife is affected by an oil spill, birds are often the most visible victims. Birds usually die more quickly and in greater numbers than other animals during an oil spill. The proceedings of the *Exxon Valdez* Oil Spill Symposium, published in 1996, indicate that 250,000 birds died during the *Exxon Valdez* oil spill, compared to 1,000 otters and 13 whales.

A bird can't fly if it gets oil on its feathers. The oil may interfere with the bird's natural waterproofing. As a result, an oiled bird can quickly die of exposure because the oil destroys the insulating ability of the bird's feathers. In trying to preen its feathers back into shape, the bird ingests oil, which is toxic to the bird and often fatal.

To have any chance of survival, oiled birds must be immediately captured and taken to a rehabilitation facility.

Oil must be removed immediately from the feathers of birds or they will not survive. Despite often prompt cleanup efforts, the bird population always suffers a big drop after an oil spill.

The longer oil remains on a bird's feathers, the lower its chances of survival.

Once in the rehabilitation center, oil is flushed from the bird's eyes and intestines. Workers also conduct a quick examination to determine if the bird has any broken bones, cuts, or other injuries. Heavily oiled birds are wiped with sorbent materials, and stomach-coating medicines may be administered to prevent the further absorption of oil into the bloodstream.

After its initial medical check and quick cleaning, the bird is placed in a curtained area and kept warm under heating lamps. This procedure helps calm the bird and reverses the effects of heat loss. According to the EPA, "Minimizing stress is critical for ensuring that captured birds survive."[21]

Once the bird is alert and its body temperature is normal, workers begin the serious cleaning. Mild detergent is gently stroked through the bird's feathers and then rinsed out. Often, a bird will need to be washed more than three times before all the oil is removed. After the bird is completely clean, it's again placed in a curtained area under heat lamps. It may start to preen and rearrange its feathers, waterproofing them and putting them back into shape.

Birds are often washed more than three times to remove all of the oil from their feathers.

A bird also has to be fed while at the rehabilitation center. As soon as a wild bird begins to feel healthy, it will usually learn to feed itself. But until the bird starts to recover, workers may have to force-feed it.

If the bird's condition remains stable, or improves, workers will let it swim. Swimming often stimulates the bird into further preening and waterproofing.

Before a bird can be released back into the wild, it has to pass several tests: The bird has to demonstrate that it can adequately preen and waterproof its feathers; it has to demonstrate that it can float; the bird has to have normal weight and muscle structure; and it must demonstrate normal feeding habits. Once a bird has passed these tests, workers gradually expose it to temperatures it will encounter outdoors. Finally, the U.S. Fish and Wildlife Service bands the bird so that it can be identified if it is found at some later date.

Even with all these efforts, most oiled birds don't survive. Prince William Sound is home to a large proportion of the entire North American population of marbled murrelets. During the *Exxon Valdez* disaster, 30 percent of the marbled murrelet population in Prince William Sound was killed. As of 1997, the U.S. government was still listing marbled murrelets as "not recovered." Harlequin ducks, common murres, and pigeon guillemots all suffered similar population declines in Prince William Sound and are also listed as not recovered.

Rescuing and cleaning furbearing mammals

Many species of marine mammals are affected by oil spills. Pinnipeds, such as harbor seals, fur seals, walruses, and sea lions, can sustain severe injuries because they spend a lot of time in the water—where the oil is—and the rest of the time hauled-out on beaches where they also often encounter oil.

Otters are also at great risk during oil spills, even more than pinnipeds. Otters spend a lot of time on the surface of the water, where their fur gets coated with petroleum and they breathe in the toxic fumes as the lighter petroleum fractions evaporate. Otters groom their fur, ingesting toxic petroleum in the process. The cleanup process for pinnipeds and otters is essentially the same, although otters require more medical efforts and have a higher mortality rate because of the toxic effects of the ingested oil.

Once an animal is captured and taken to the rehabilitation center, it is checked for signs of hypothermia (low

Recovery is a long process and several tests must be given to an animal before it is deemed healthy enough to return to the wild.

body temperature), dehydration (low body fluid levels), and physical injuries. Otters are particularly at risk for hypothermia because they do not have an insulating layer of blubber as pinnipeds do. Workers stabilize the animal's body temperature and fluid levels and prepare it for cleaning by administering a mild sedative.

Two people work on each animal. One person holds and restrains the animal while the other washes it with a mild detergent. After washing, otters are hand-rubbed with towels and then dried with hair dryers. Once clean, the otter will groom and waterproof itself. It takes about seven days of grooming for an otter's coat to regain its natural water-repellency. After washing, pinnipeds are dried off a bit, then kept in pens, where they continue to dry naturally.

Once an animal is clean, workers may keep it in captivity for days, weeks, or even months, until its health improves. During this recuperation period, the animal's body temperature, eating habits, and general health are monitored. Workers feed the animals a variety of their favorite foods, keep them warm, and give them medicine as needed.

As an animal's health improves, workers start exposing it to cooler temperatures and allow it to try swimming.

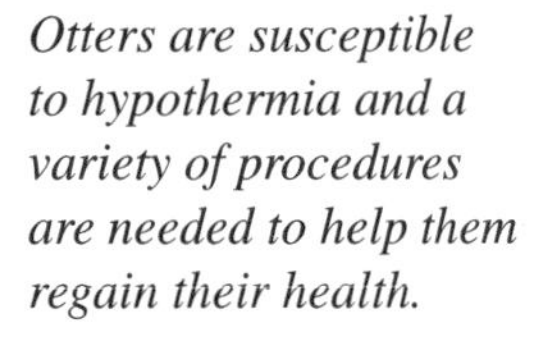

Otters are susceptible to hypothermia and a variety of procedures are needed to help them regain their health.

Slowly, it is reintroduced to the conditions of its natural habitat.

Sometimes an animal is well, but its natural habitat is still fouled with oil. In such cases, workers will either keep the animal until its habitat is cleaned up, or they will try and release it into a similar, but oil-free, habitat. Releasing animals into a different habitat doesn't always work. Because otters are territorial, they will often try to return to their original habitat, even if that area is fouled with oil. To combat this problem, released otters are tagged with tracking devices. Rescue workers can then tell where they go after they're released. If they attempt to return to an oiled area, workers can retrieve them and try again. Tracking devices also help scientists monitor an animal's health after an oil spill.

Other marine mammals

Whales are also at risk in oil spills, although it is more difficult to know how severely they are affected because they are harder to find and track than other marine mammals. Because whales don't have fur and spend more time in deep water than do other marine mammals, scientists used to think they were less immediately affected. Yet even with these advantages, whales are still injured and killed by oil spills. Moreover, a whale's food supply is often wiped out by oil spills. In fact, scientists rarely learn how many whales are injured or killed during oil spills because most injured or dead whales simply sink to the bottom of the ocean. Even when an injured whale is found, there is nothing workers can do. There is no way to treat a whale in the wild, and there is no way to keep a large whale in captivity.

During the *Exxon Valdez* disaster, scientists had their first opportunity to discover what really happens to whales during an oil spill. Prince William Sound is home to a resident pod of killer whales that have been extensively studied for at least twenty years. The pod contained thirty-six members before the spill. During the spill the pod lost thirteen of its members, and it produced no young during the first two years following the spill. Prior to the *Exxon Valdez* disaster, scientists had thought that whales fared better than other marine mammals during oil spills simply because they had no basis of comparison and no other information.

According to the Oil Spill Public Information Center's "1997 Status Report" on the *Exxon Valdez* oil spill, "the losses (of the Prince William Sound pod of killer whales) far exceed normal rates documented over 20 years of study for this and other killer whale groups in the north Pacific."[22] Most scientists who have studied Prince William Sound believe that the declines in this pod are directly attributable to the effects of the *Exxon Valdez* oil spill.

Tracking the long-term effects

With every oil spill, scientists learn more about the long-term effects. Not every animal injured in an oil spill is given a tracking device, but every animal that is injured, treated, and released is given an identification tag. In fact, during an oil spill every animal that is captured is given a tag, even if it has no injuries. Through this tagging system, scientists hope to understand the long-term effects of oil spills on animal and marine life. If an animal with a tag is sighted, its location is reported to scientists, who can often identify the animal and determine how the animal is doing in the wild. If a tagged animal dies and is found, scientists are notified and can determine the cause of death.

Because this identification and tracking system was used during the *Exxon Valdez* oil spill, scientists have already learned a great deal about the long-term effects of an oil spill on animal life. They have learned that many animals that have been oiled do not survive, no matter how well they were cleaned and medicated.

For example, as part of the cleanup efforts following the *Exxon Valdez* spill, 357 otters were captured and treated. More than one-third of these animals died in captivity. Others survived but were not healthy enough to be released into the wild. Twenty-four adult otters and 13 pups were

Although extensive treatment was given to otters affected by the Exxon Valdez *spill, after nine months, nearly half of the 197 released into the wild were dead or presumed dead.*

transferred to aquariums around the world because they were too ill or too young to be released.

After the spill 197 otters were released back into Prince William Sound. Most of the released otters were fitted with tracking devices. In less than nine months, half of these animals were dead, or missing and presumed dead. Autopsies on the recovered animals revealed that most had suffered extensive kidney, lung, and liver damage.

The cost of an oil spill

Although much of the damage an oil spill does cannot be measured in pure dollar amounts, a great deal of it can. During the *Exxon Valdez* oil spill, for example, the Exxon Corporation spent more than eighty thousand dollars on each rescued otter.

Additionally, according to figures published by the state of Alaska, between March 1989 (when the spill occurred) and August 1991, Exxon spent more than $2.1 billion for cleanup activities they did themselves as well as reimbursements to federal, state, and local governments for their expenses in responding to the spill.

Throughout the world, the public is becoming aware that any oil spill is expensive and harmful, that no spill can be completely cleaned up, and that the short-term and long-term effects on plants and wildlife are devastating. Because no cleanup method is 100 percent effective, scientists feel that in the future the emphasis must be on the prevention of oil spills. The potential for oil spills increases as more oil is moved about the globe. Prevention is the only way to diminish the risk.

6

Prevention and the Future

AS TANKERS AND pipelines move more oil around the world, the potential for disastrous oil spills increases. Most scientists believe that the future emphasis should be on prevention of oil spills. According to the EPA, "We can best avoid the environmental and economic effects of oil spills by preventing and containing them in the first place."[23]

After the *Exxon Valdez* disaster, many people felt that oil companies were not really prepared to handle a large oil disaster. They felt that petroleum companies should be better prepared and ready to respond immediately to any oil-spill emergency. Shortly after the *Exxon Valdez* oil spill, fourteen oil companies responded to public pressure by agreeing to set up five different centers, each located in a different region of the United States. The purpose of these centers was twofold: to train personnel and to have equipment ready so that in the event of an oil spill emergency cleanup measures could begin immediately. Everything would be ready and waiting; there would be no lag time.

Although these centers do not prevent oil spills, they do help to prevent some of the environmental damage that occurs during such spills. Time is of the essence for preventing damage, so the sooner cleanup workers can reach the site of an oil spill with equipment, the better they can prevent oil from reaching shorelines and other ecologically sensitive areas.

In the early nineties, public outrage over the Exxon Valdez *spill promoted new laws that imposed massive penalties on companies for large oil spills. Consequently, the number of oil spills has been greatly reduced today.*

Public anger over the *Exxon Valdez* oil spill also led the U.S. Congress to pass the Oil Pollution Act of 1990. This law required that oil companies have better emergency preparations in place to deal with spills. The new law also considerably raised the monetary penalties oil companies would face should oil spills occur.

Financial motivation for prevention

Until 1990 no strong financial incentives existed to encourage oil companies to avoid spills. Consequently, prevention and cleanup were not a priority. Since the passage of the Oil Pollution Act, an oil spill can cost a petroleum company hundreds of millions of dollars.

The heavy financial penalties oil companies now face appear to be having an effect on the amount of oil spilled. For the year 1991, the amount of oil spilled into the environment was the lowest amount since 1978. The amount of oil spilled in the environment has declined every year since the passage of the Oil Pollution Act, and no major oil-spill disaster has occurred in U.S. waters or on land.

Safer oil tankers

Stiff financial penalties clearly provide incentive for oil companies to prevent spills. The Oil Pollution Act of 1990 and the international maritime regulations that have followed encourage prevention not only by creating incentives but by requiring safer tankers. By the year 2020 nearly all vessels used to transport oil will be required to have double hulls. Double bottoms will also be required on tankers. Of the more than three thousand tankers operating in the world today, only about five hundred of them have double bottoms. Double hulls and double bottoms will lessen the chances of oil spills caused by punctures, which currently cause 70 percent of all maritime oil-spill disasters.

According to a report issued by the National Academy of Sciences,

> About 10 percent of oil-transporting vessels had double hulls as of 1994, and many new ones will enter service within the next few years as the industry complies with U.S. and international requirements. Replacing all single-hull tankers with correctly designed double-hull vessels could prevent a great number of spills attributed to collisions and groundings, and reduce by as much as two-thirds the total volume of oil spilled by such accidents.[24]

Other people are not so sure that double-hulled and double-bottomed vessels will solve the problem. As Seaman Joseph Gross points out in an article in the *Ethical Spectacle*,

> The void spaces on ships are not full of inert gas, they are full of air, which contains the oxygen required for combustion. Consider the following scenario: A large tank vessel develops a crack in one of its tanks, allowing oil and oil vapor to leak into the void space, which is inspected only periodically. Suppose that over a period of weeks or months, a considerable amount of vapor may accumulate in this space. The tanker runs aground, creating sparks where the hull is penetrated. Clearly, this is the recipe for a devastating explosion, fire, and oil spill. The void space is now a tremendous liability.[25]

Despite Gross's view, the majority of scientists are in favor of double-hulled and double-bottomed tankers.

Controlling human error

Human error can be a contributing factor in an oil spill. Human error can involve everything from poor judgment on the captain's part to an overworked crew, improper training, or inadequate tug and navigation services. Almost anything short of a severe storm can be attributed to human error. For example, the *Exxon Valdez* oil spill in Prince William Sound, Alaska, was blamed entirely on human error. According to the Oil Spill Public Information Center,

> The National Transportation Safety Board determined that the probable cause of the grounding of the Exxon Valdez was the failure of the third mate to properly maneuver the vessel because of fatigue and excessive workload; the failure of the master to provide a proper navigation watch because of impairment from alcohol; the failure of the Exxon Shipping Company to provide a fit master and a rested and sufficient crew for the Exxon Valdez; the lack of an effective vessel traffic service because of inadequate equipment and manning

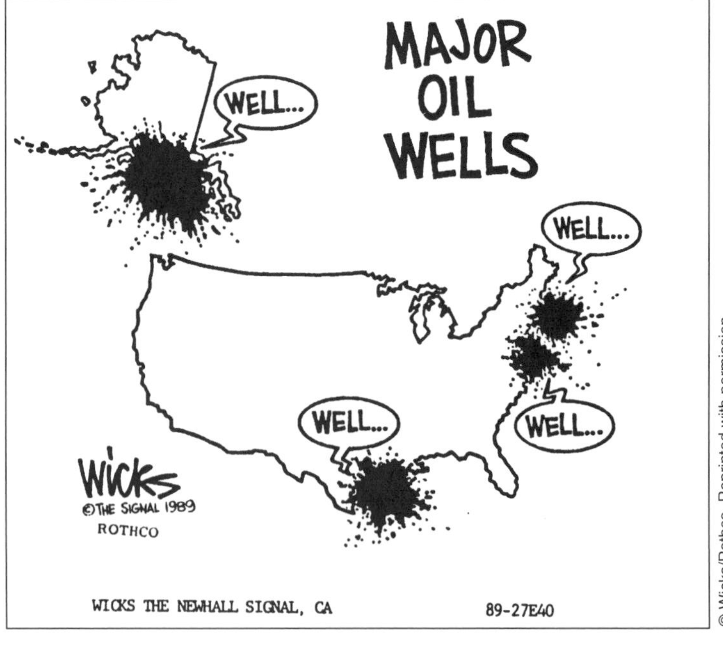

> levels, inadequate personnel training, and deficient management oversight; and the lack of effective pilotage services.[26]

In response to disasters like the *Exxon Valdez* spill, petroleum companies in the United States are now making sure that their tanker crews have better training, more supervision, and adequate numbers of workers for each shift. Most experts believe that better training and shorter, less stressful shifts will lower the incidence of human error.

Around the globe, better navigational aids continue to be developed. In 1996, for example, an advanced navigational system called the Physical Oceanographic Real-Time System (PORTS) was installed to help vessels safely navigate in San Francisco Bay. According to the National Oceanic and Atmospheric Administration (NOAA),

> With PORTS, captains and pilots of large tanker ships have access to the real-time nautical data required to plan their arrivals and departures to maximize the efficiency of their time in port. PORTS measures currents, water levels and other physical conditions on a 24-hour basis to allow ship captains and pilots to use the Bay's channels to their fullest extent. PORTS is a vital step in preventing oil spills; it removes much of the potential for human error for large oil tanker traffic.[27]

Another approach to preventing spills—or at least minimizing their effects—involves greater regulation of shipping. One proposed regulation calls for all large tankers to remain one hundred miles offshore. Rather than large tankers docking at a port, smaller vessels would lighter the oil off the tankers and take it to shore. This way, should an oil spill occur, the chances that the spill would reach and damage beaches and marshes would be lower. Moreover, by staying far out to sea, ships—in theory, at least—should be less likely to run into reefs or other obstructions.

More domestic oil

Many people advocate using more domestic oil as a way to lower the chance of an oil spill. Executives of many U.S. oil companies argue that if more areas in the United States were opened to exploration and drilling, we wouldn't need to import as much oil from foreign countries. Since the

United States is the major consumer of oil in the world, less oil would be traveling around the globe by tanker, and the chances of a spill would decrease. The possibility of land-based spills would still exist, but the vast majority of oil spills involve oceangoing ships, so the proponents of using more domestic oil feel that this is a viable way to decrease the chances of a disastrous oil spill occurring in the United States.

Many of the areas that U.S. oil companies want to drill in are currently closed because they are considered environmentally sensitive or pristine. In August 1998, however, the U.S. government agreed to allow oil companies to drill in specific areas of the Alaskan North Slope wilderness. Although oil companies applauded the decision, others were dismayed.

In a *New York Times* article, William H. Meadows, president of the Wilderness Society, an environmental organization based in Washington, D.C., stated that "this is a terribly shortsighted decision. This reserve contains some world-class wildlife habitat, vital to grizzly bears, caribou, and millions of shorebirds, ducks, and geese."[28]

Further drilling and exploitation of U.S. oil might seem like a logical solution to lowering the risk of tanker oil spills, but the argument is not entirely true. Pipeline accidents could still occur; the *Exxon Valdez* oil spill occurred in U.S. waters, with a U.S. tanker that was carrying U.S. oil—oil drilled from Alaska's North Slope.

Prevention of pipeline accidents

The United States averages fewer than one hundred pipeline accidents per year; and, so far, most of them have been less than one thousand gallons. As far as spills go, pipelines in the United States have a good record. Short of sabotage or deliberate attack, pipelines are a relatively safe way to transport oil.

Overall, pipelines are becoming the preferred method of transporting oil, primarily because they are considered safer than tankers. In October 1998 the United States gave Turkey $823,000 toward the construction of a new pipeline

In the future, oil pipelines could greatly reduce the number of oil accidents if properly maintained.

that will run across that country. The discovery of huge amounts of oil reserves under the Caspian Sea has led to increased oil production in the area, and Turkey does not want oil-tanker traffic in the Bosporous Strait or Caspian and Black Seas. As Imren Aykut, the Turkish minister of environment, announced, "No country has the right to endanger the lives of 10 million people just because it wants to sell oil. We will never allow our sea lanes to be turned into oil pipelines."[29] American oil companies, such as Exxon, Amoco, and Pennzoil, are expected to also contribute money to the construction of Turkey's oil pipeline.

If properly maintained, pipelines are a safe way to transport oil. Ongoing maintenance of a pipeline, though, is critical to the prevention of oil spills. Corrosion is a large problem that affects pipelines. Due to the hydrocarbons and other components of oil, the petroleum itself can corrode the inside of the pipeline. Likewise, normal weather conditions take their toll on the exterior of the pipeline. Regular maintenance of the pipeline requires routine checks of the whole pipeline and regular replacement of sections that show any sign of corrosion.

Pipeline valves must also be maintained. During construction, valves are placed at consistent intervals along the pipeline so that, if a break occurs, the flow of oil through the damaged section of pipe can be stopped immediately by closing the valve. To ensure that pipelines stay safe, workers have to check and replace the valves on a regular basis.

People, not oil companies, preventing spills

Correctly recycling oil is another way to stop oil from entering our environment. More than half of America's motorists change their own motor oil. These motorists generate more than 200 million gallons of used oil each year—enough to fill five supertankers. Frequently, this oil is disposed of improperly—dumped into garbage cans, down sewers, or in backyards. More than 150 million gallons of used oil ends up in our environment every year, contaminating soil and water.

If oil is recycled correctly, at a collection and recycling station, it can be re-refined into industrial fuel and lubricants. Based on a pilot program developed by the state of Florida, sixteen other states have adopted official programs to recycle used oil. As of 1998 twelve more states were considering such programs to help keep oil out of the environment. The programs involve setting up oil recycling centers and then advertising to let the general public know of their existence. The centers then deliver the used oil to refineries where it can be turned into other useful products.

These prevention and recycling programs appear to be working. The American Petroleum Institute reports that between 1991 and 1998 nearly 81 million gallons of used motor oil had been collected by recycling centers around the United States. Much remains to be done, though.

Many people are trying to change their lifestyles to help conserve and recycle all our natural resources, including oil. As industries and nations learn to use resources like oil more efficiently, perhaps the world's dependence on oil will lessen, oil-tanker traffic will taper off, and the chances of a disastrous oil spill fouling our water, beaches, bays, and marshes will also diminish. This is one strategy that will assure future generations a clean and healthy environment.

Notes

Chapter 1: Oil from Rocks

1. Norman J. Hyne, *Nontechnical Guide to Petroleum Geology, Exploration, Drilling, and Production*. Tulsa, OK: PennWell Books, 1995, p. 1.

2. Chevron Corporation, "How the Oil Industry Developed," available from http://www.chevroncars.com/know/oil/index.html.

3. *Houston News Today* Online, "Texas Monthly Oil and Gas Statistics," available from http://www.houstontoday.com/_btmain/oilandgasstatistics/index.html.

Chapter 2: Oil in the Environment

4. National Oceanic and Atmospheric Administration, "Managing Spills of Oil and Chemical Materials—Introduction," *State of the Coast Report,* available from http://state_of_coast.noaa.gov/bulletins/html/hms_15/intro.html.

5. Forest Gray, *Petroleum Production in Nontechnical Language.* Tulsa, OK: PennWell Books, 1995, p. 58.

6. Cutter Information Corporation, "White Papers," *Oil Spill Intelligence Report,* available from http://www.cutter.com/osir/osirpape.htm.

7. Wesley Marx, *Oil Spill.* San Francisco: Sierra Club Books, 1971, p. 49.

8. National Academy of Sciences, National Research Council, "Oil Spill Risks from Tank Vessel Lightering," project identification number MBXX-J-97-05-A, available from http://www4.nas.edu/webcr.nsf.

Chapter 3: Oil Spills

9. U.S. Environmental Protection Agency (EPA), Oil Spill Program, "The Fate of Spilled Oil," available from http://www.epa.gov/oilspill/oilfate.htm.

10. Australian Maritime Safety Authority, Marine Environment Protection, "What Happens to Spilled Oil?" available from http://www.amsa.gov.au/me/edu/happens.htm.
11. U.S. EPA, Oil Spill Program, "The Fate of Spilled Oil."
12. U.S. EPA, Oil Spill Program, "The Fate of Spilled Oil."

Chapter 4: Cleaning Up the Environment

13. U.S. EPA, Oil Spill Program, "Booms," available from http://www.epa.gov/oilspill/booms.htm.
14. U.S. EPA, Oil Spill Program, "Booms."
15. U.S. EPA, Oil Spill Program, "Sorbents," available from http://www.epa.gov/oilspill/sorbents.htm.
16. U.S. EPA, Oil Spill Program, " Dispersing Agents," available from http://www.epa.gov/oilspill/ disperse.htm.
17. U.S. EPA, Oil Spill Program, " Response Techniques," available from http://www.epa.gov/oilspill/oiltech.htm.

Chapter 5: Cleaning Up Wildlife

18. Laurence Pringle, *Oil Spills: Damage, Recovery, and Prevention.* New York, Morrow Junior Books, 1993, p. 28.
19. U.S. EPA, Oil Spill Program, "Rescuing Wildlife," available from http://www.epa.gov/oilspill/rescue.htm.
20. U.S. EPA, Oil Spill Program, "Rescuing Wildlife."
21. U.S. EPA, Oil Spill Program, "Rescuing Wildlife."
22. Oil Spill Public Information Center, "1997 Status Report," section on killer whales, available from http://www.alaska.net/~ospic/recover. html.

Chapter 6: Prevention and the Future

23. U.S. EPA, Oil Spill Program, "Preventing Spills," available from http://www.epa.gov/oilspill/prevent.htm.
24. National Academy of Sciences, National Research Council, "Double-Hull Vessels Could Significantly Reduce Oil Spills, but New Design Standards Are Needed," November 6, 1997, available from http://www2.nas.edu/whatsnew/27de.html.
25. Joseph Gross, "Tankers and Spills," *Ethical Spectacle,* January 1996, available from http://www.spectacle.org/196/gross.html.

26. Oil Spill Public Information Center, "Exxon Valdez Oil Spill Frequently Asked Questions," available from http://www.alaska.net/~ospic/faq.html.
27. National Oceanic and Atmospheric Administration, "Managing Spills of Oil and Chemical Materials—Case Studies," *State of the Coast Report*, available from http://state_of_coast.noaa.gov/bulletins/html/hms_15/case.html.
28. Quoted in John H. Cushman Jr., "In Compromise, U.S. Opens Part of Alaska North Slope to Drilling," *New York Times*, August 6, 1998, available from http://www.nytimes.com/98/08/06.html.
29. Quoted in Stephen Kinzer, "Fearless Turks' Big Fear? Oil Tankers," *New York Times*, October 24, 1998, available from http://www.nytimes.com/98/10/24.html.

Glossary

asphalt: One of the heaviest petroleum fractions. Asphalt is what remains after all other fractions have been boiled off.

biodegradation: The natural process of bacteria and microorganisms consuming and helping to destroy an oil spill.

bitumen: A fraction of petroleum that is found in tar sands. To be usable, bitumen must be refined.

booms: A fencelike structure that lies on the surface of the water and helps to keep oil from spreading.

burning: The process of igniting and burning an oil spill. Burning is always done under controlled conditions, and, to work well, the spill must be fresh.

cracking: The process of refining a petroleum fraction into further chemicals and compounds.

crude: The term for petroleum in its natural state. Crude oil refers to petroleum that has not been refined.

dispersants: Chemical compounds that, when applied to an oil spill, will cause it to break down into smaller particles and be absorbed into water. Dispersants are controversial and rarely used in the United States.

emulsification: The process by which oil and water are whipped together, creating a frothy puddinglike mixture.

fertilization: The process of "feeding" natural bacteria and microorganisms to speed biodegradation.

fractions: The different parts of crude oil, or natural petroleum. When crude oil is boiled, it produces different components, called fractions.

gelling agents: Chemical compounds that are added to an oil spill and cause the oil to gel and clump together making it easier to suction and clean up.

hydrocarbons: The organic basis of petroleum. Hydrocarbons are formed from organic matter and have unique identifying characteristics.

lightering: The process of moving oil from one tanker onto another.

oxidation: The process of air mixing with an oil spill. Oxidation aids evaporation and also creates deadly water-soluble chemicals.

petroleum: Oil in almost any form—crude, gasoline, kerosene, tar, asphalt, or bitumen—is referred to as petroleum.

sedimentary rock: The type of rock formed when layers of organic debris and sediment, or sand, are compressed together.

sedimentation: The process by which oil mixes with the natural, often microscopic, sediments that are in water. As the oil clumps with the tiny bits of clay and sand that are dissolved in the water, it becomes heavier and sinks to the bottom.

seeding: The process of adding foreign nutrients and other growth stimulators to an oil spill in an effort to make bacteria and microorganisms biodegrade faster.

skimmer: A piece of equipment that sucks and skims oil off the surface of the water.

sorbents: Materials that will attract and soak up spilled oil. Sorbents include straw, sawdust, and other materials.

spreading: The first step of an oil spill on water. When oil encounters water, it will spread out over the surface, causing an oily sheen.

tar: One of the heaviest fractions of petroleum, tar is similar to asphalt and is used to pave roads and waterproof roofs.

weathering: The process by which wind, wave action, currents, water temperature, and microorganisms work together to begin neutralizing spilled oil.

Organizations to Contact

Alaska Resources Library and Information Services
(ARLIS)
3150 C St., Suite 100
Anchorage, AK 99503
(907) 27-ARLIS

ARLIS is a partnership of eight Anchorage libraries that offer complete information pertaining to oil spills, their cleanup, and their aftermath.

American Petroleum Institute (API)
1220 L St. NW
Washington, DC 20005-4070
(202) 682-8000
web address: http://www.api.org

A trade organization for the petroleum industry, API offers many fact sheets and useful information about oil.

National Oceanic and Atmospheric Administration (NOAA)

web address: http://www.noaa.gov/

NOAA is responsible for overseeing fish and other aquatic wildlife as well as the quality of U.S. waters. Consequently, much that happens during an oil spill is within its jurisdiction. NOAA publishes many fact sheets and provides detailed information on many subjects. It has many divisions and offices around the country and is best reached through its website.

Oil Spill Public Information Center
645 G St.
Anchorage, AK 99501-3451
(800) 283-7745 (outside Alaska)
(800) 478-7745 (inside Alaska)
fax: (907) 265-9359
web address: http://www.alaska.net/~ospic/

The center offers comprehensive and complete information about oil spills, particularly the *Exxon Valdez* oil spill.

U.S. Environmental Protection Agency (EPA)
web address: http://www.epa.gov/

The EPA is responsible for monitoring the cleanup of all oil spills. The EPA maintains different offices around the country and is best reached through its website.

Suggestions for Further Reading

Art Davidson, *In the Wake of the* Exxon Valdez. San Francisco: Sierra Club Books, 1990. A dramatic account of the crash of the *Exxon Valdez* in Prince William Sound, Alaska.

Christopher Lampton, *Oil Spill*. Brookfield, CT: Millbrook Press, 1992. This book offers a quick, basic overview of what happens during an oil spill.

Steve Parker, *The Earth and How It Works*. London: Dorling Kindersley, 1989. A lower-reading-level book that gives an interesting account of how petroleum is formed and where it is found.

Laurence Pringle, *Oil Spills: Damage, Recovery, and Prevention*. New York: Morrow Junior Books, 1993. Although written for a slightly younger audience, this book gives a highly informative account of all phases of an oil spill as well as some interesting suggestions for conservation.

Anne Gruner Schlumberger, *The Schlumberger Adventure*. New York: Arco, 1982. Originally published in French, it's an interesting account of the early days of the oil industry.

Page Spencer, *White Silk and Black Tar: A Journal of the Alaska Oil Spill*. Minneapolis: Bergamot Books, 1990. An interesting journal written by a scientist who works for the National Park Service in Anchorage. It chronicles many of the events involved with the *Exxon Valdez* oil spill.

Darlene R. Stille, *Oil Spills*. Chicago: Childrens Press, 1991. Although this book is intended for younger readers, it gives a good overview of an oil spill.

Susanna Van Rose, *Earth*. London: Dorling Kindersley, 1994. Although aimed at a younger audience, this comprehensively illustrated book offers useful information on the formation of petroleum.

Works Consulted

Books

Parke A. Dickey, *Petroleum Development Geology*. Tulsa, OK: PennWell Books, 1979. A somewhat technical but very informative book on searching and drilling for oil.

Forest Gray, *Petroleum Production in Nontechnical Language*. Tulsa, OK: PennWell Books, 1995. A very readable and informative book on the production and refining of oil.

Norman J. Hyne, *Nontechnical Guide to Petroleum Geology, Exploration, Drilling, and Production*. Tulsa, OK: PennWell Books, 1995. This book provides a clear explanation of petroleum geology and production.

Robert D. Langenkamp, *Oil Business Fundamentals*. Tulsa, OK: PennWell Books, 1982. An informative account of the production and marketing aspects of the petroleum industry.

Wesley Marx, *Oil Spill*. San Francisco: Sierra Club Books, 1971. A highly readable book that portrays the environmental consequences of oil spills.

Jeffrey Potter, *Disaster by Oil. Oil Spills: Why They Happen, What They Do, How We Can End Them*. New York: Macmillan, 1973. A readable account of some of the major oil spills in the twentieth century.

Periodicals

Associated Press, "Oil Pipeline Blows Up in Nigeria, Killing 250," *New York Times*, October 19, 1998.

John H. Cushman Jr., "In Compromise, U.S. Opens Part of Alaska North Slope to Drilling," *New York Times*, August 6, 1998, available from http://www.nytimes.com/98/08/06.html.

Suzanne Daley, "Why This Pampered Paradise? It's Oil, Stupid!" *New York Times*, June 25, 1998.

Joseph Gross, "Tankers and Spills," *Ethical Spectacle,* January 1996, available from http://www.spectacle.org/196/gross.html.

Stephen Kinzer, "Fearless Turks' Big Fear? Oil Tankers," *New York Times*, October 24, 1998, available from http://www.nytimes.com/98/10/24.html.

——, "U.S., Pushing Its Route for Pipeline, Aids Turkey," *New York Times*, October 22, 1998.

Serge F. Kovaleski, "Bombs Close Colombian Oil Pipeline," *Washington Post*, June 24, 1998.

Diana Jean Schemo, "Oil Pipeline Blast and Fire in Colombia Kill 45, Mostly Villagers; Rebels Are Blamed," *New York Times*, October 19, 1998.

Websites

American Petroleum Institute, "Collecting and Recycling Used Motor Oil," *Strategies for Today's Environmental Partnership,* available from http://www.api.org/pasp/step/usedoil.htm.

——, "Facts About Oil," available from http://www.api.org/edu/factsoil.htm.

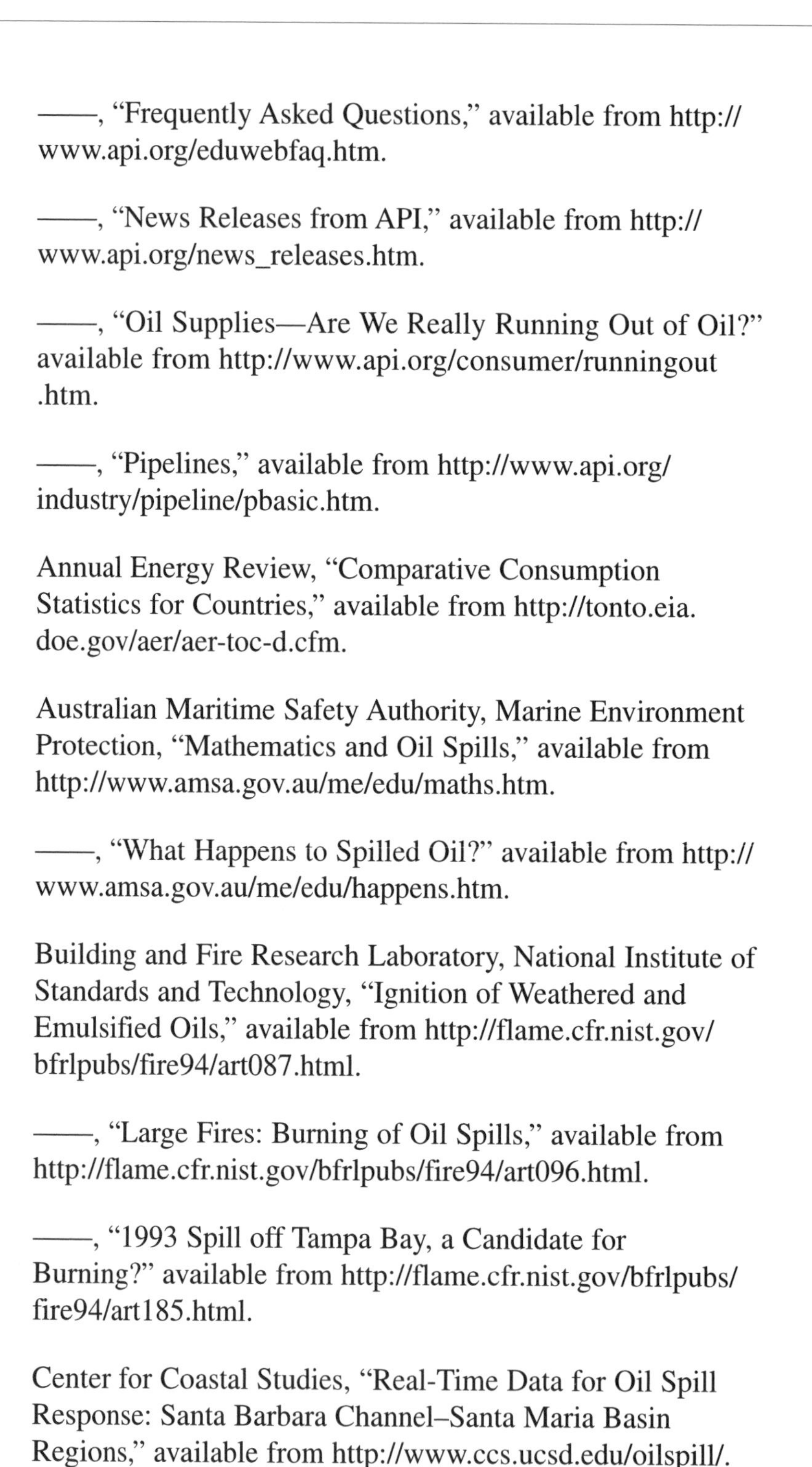

——, "Frequently Asked Questions," available from http://www.api.org/eduwebfaq.htm.

——, "News Releases from API," available from http://www.api.org/news_releases.htm.

——, "Oil Supplies—Are We Really Running Out of Oil?" available from http://www.api.org/consumer/runningout.htm.

——, "Pipelines," available from http://www.api.org/industry/pipeline/pbasic.htm.

Annual Energy Review, "Comparative Consumption Statistics for Countries," available from http://tonto.eia.doe.gov/aer/aer-toc-d.cfm.

Australian Maritime Safety Authority, Marine Environment Protection, "Mathematics and Oil Spills," available from http://www.amsa.gov.au/me/edu/maths.htm.

——, "What Happens to Spilled Oil?" available from http://www.amsa.gov.au/me/edu/happens.htm.

Building and Fire Research Laboratory, National Institute of Standards and Technology, "Ignition of Weathered and Emulsified Oils," available from http://flame.cfr.nist.gov/bfrlpubs/fire94/art087.html.

——, "Large Fires: Burning of Oil Spills," available from http://flame.cfr.nist.gov/bfrlpubs/fire94/art096.html.

——, "1993 Spill off Tampa Bay, a Candidate for Burning?" available from http://flame.cfr.nist.gov/bfrlpubs/fire94/art185.html.

Center for Coastal Studies, "Real-Time Data for Oil Spill Response: Santa Barbara Channel–Santa Maria Basin Regions," available from http://www.ccs.ucsd.edu/oilspill/.

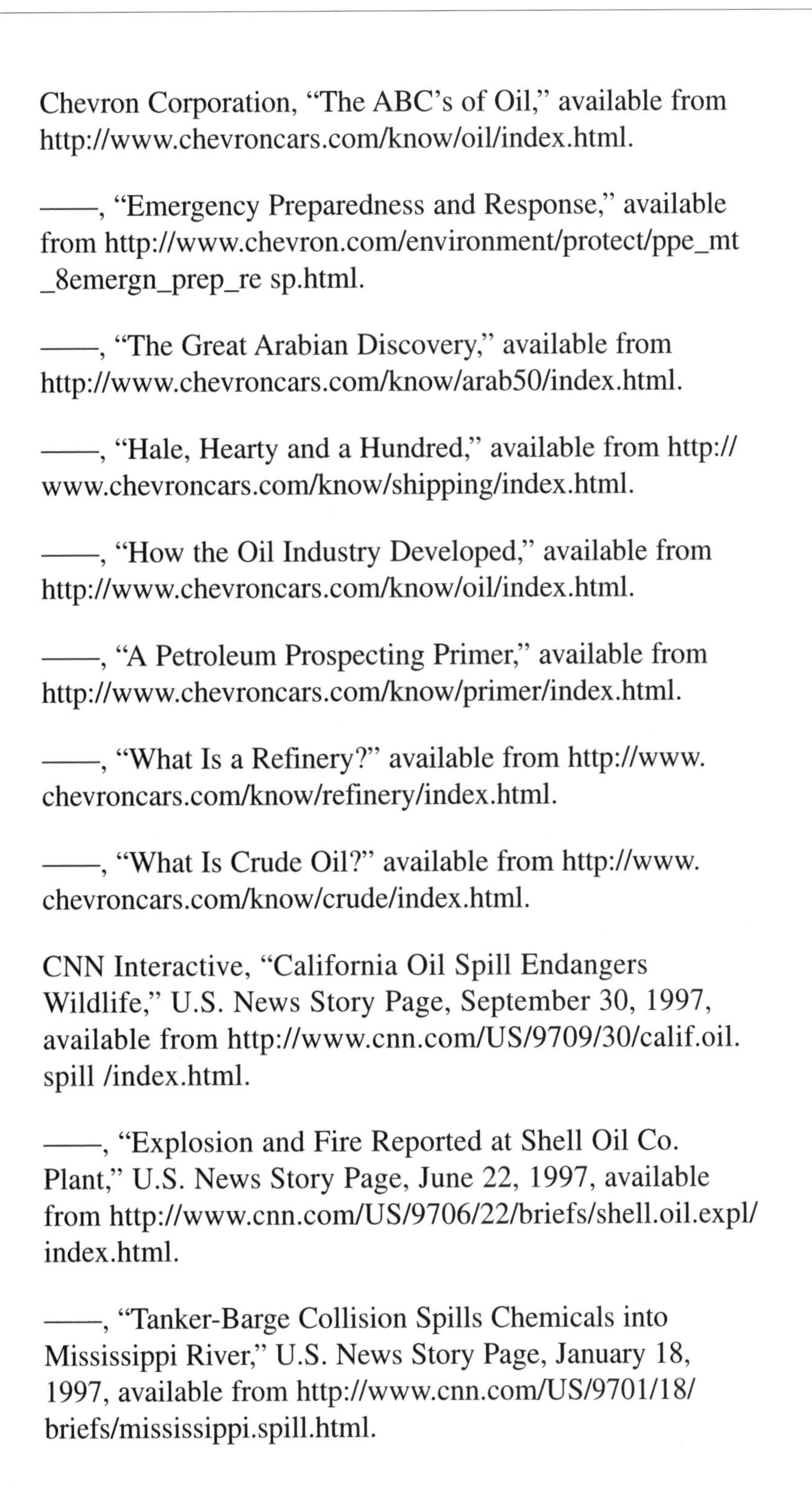

Chevron Corporation, “The ABC’s of Oil,” available from http://www.chevroncars.com/know/oil/index.html.

——, “Emergency Preparedness and Response,” available from http://www.chevron.com/environment/protect/ppe_mt _8emergn_prep_re sp.html.

——, “The Great Arabian Discovery,” available from http://www.chevroncars.com/know/arab50/index.html.

——, “Hale, Hearty and a Hundred,” available from http://www.chevroncars.com/know/shipping/index.html.

——, “How the Oil Industry Developed,” available from http://www.chevroncars.com/know/oil/index.html.

——, “A Petroleum Prospecting Primer,” available from http://www.chevroncars.com/know/primer/index.html.

——, “What Is a Refinery?” available from http://www.chevroncars.com/know/refinery/index.html.

——, “What Is Crude Oil?” available from http://www.chevroncars.com/know/crude/index.html.

CNN Interactive, “California Oil Spill Endangers Wildlife,” U.S. News Story Page, September 30, 1997, available from http://www.cnn.com/US/9709/30/calif.oil.spill /index.html.

——, “Explosion and Fire Reported at Shell Oil Co. Plant,” U.S. News Story Page, June 22, 1997, available from http://www.cnn.com/US/9706/22/briefs/shell.oil.expl/index.html.

——, “Tanker-Barge Collision Spills Chemicals into Mississippi River,” U.S. News Story Page, January 18, 1997, available from http://www.cnn.com/US/9701/18/briefs/mississippi.spill.html.

Compton's Encyclopedia Online, "Abadan, Iran," available from http://comptons2.aol.com/encyclopedia/ARTICLES/00007_A.html.

Cutter Information Corporation, "White Papers," *Oil Spill Intelligence Report,* available from http://www.cutter.com/osir/osirpape.htm.

Exxon Corporation, "Applying Biological Technology to Help Clean Sites," section 8: developing technology, available from http://www.exxon.com/exxoncorp/.

——, "Protecting the Environment," available from http://www.exxon.com/exxoncorp/.

Houston News Today Online, "Texas Monthly Oil and Gas Statistics," available from http://www.houstontoday.com/_btmain/oilandgasstatistics/index.html.

National Academy of Sciences, National Research Council, "Double-Hull Vessels Could Significantly Reduce Oil Spills, but New Design Standards Are Needed," November 6, 1997, available from http://www2.nas.edu/whatsnew/27de.html.

National Oceanic and Atmospheric Administration, "Managing Spills of Oil and Chemical Materials—Case Studies," *State of the Coast Report,* available from http://state_of_coast.noaa.gov/bulletins/html/hms_15/case.html.

——, "Managing Spills of Oil and Chemical Materials—Introduction," *State of the Coast Report,* available from http://state_of_ coast.noaa.gov/bulletins/html/hms_15/intro.html.

——, "Oil Spill Risks from Tank Vessel Lightering," project identification number MBXX-J-97-05-A, available from http://www4.nas.edu/webcr.nsf/.

Natural Gas Week, "Hydraulic Fracturing," available from http://www.erols.com/clegates/FRACT.HTM.

——, "The Inorganic Petroleum Formation Theory," available from http://www.naturalgas.org/INORG.HTM.

Offshore Data Services, Inc., Online, "Weekly Mobile Offshore Rig Count," available from http://www.offshore-data.com/rigcount.html.

Oil Spill Public Information Center, "Civil and Criminal Penalties," available from http://www.alaska.net/~ospic/settle.html.

——, "Exxon Valdez Oil Spill Frequently Asked Questions," available from http://www.alaska.net/~ospic/faq.html.

——, "1995 Status Report," available from http://www.alaska.net/~ospic/status.html.

——, "1997 Status Report," section on killer whales, available from http://www.alaska.net/~ospic/recover.html.

——, "Oil Spill Public Information Center," available from http://www.alaska.net:80/~ospic/index.html.

——, "Recovery Status," available from http://www.alaska.net/~ospic/recover.html.

——, "Research, Monitoring, Restoration," available from http://www.alaska.net/~ospic/research.html.

——, "What Happened on March 24, 1989," available from http://www.alaska.net/~ospic/rpln.html.

Strategies of Oil Spill Response, "Response Strategies and Considerations," Maritime Home Page, available from http://www.webcom.com/~maritime/response/respfr.html.

Texas General Land Office, "Containment and Cleanup," available from http://www.glo.state.tx.us/oilspill/292_report/cleanup.html.

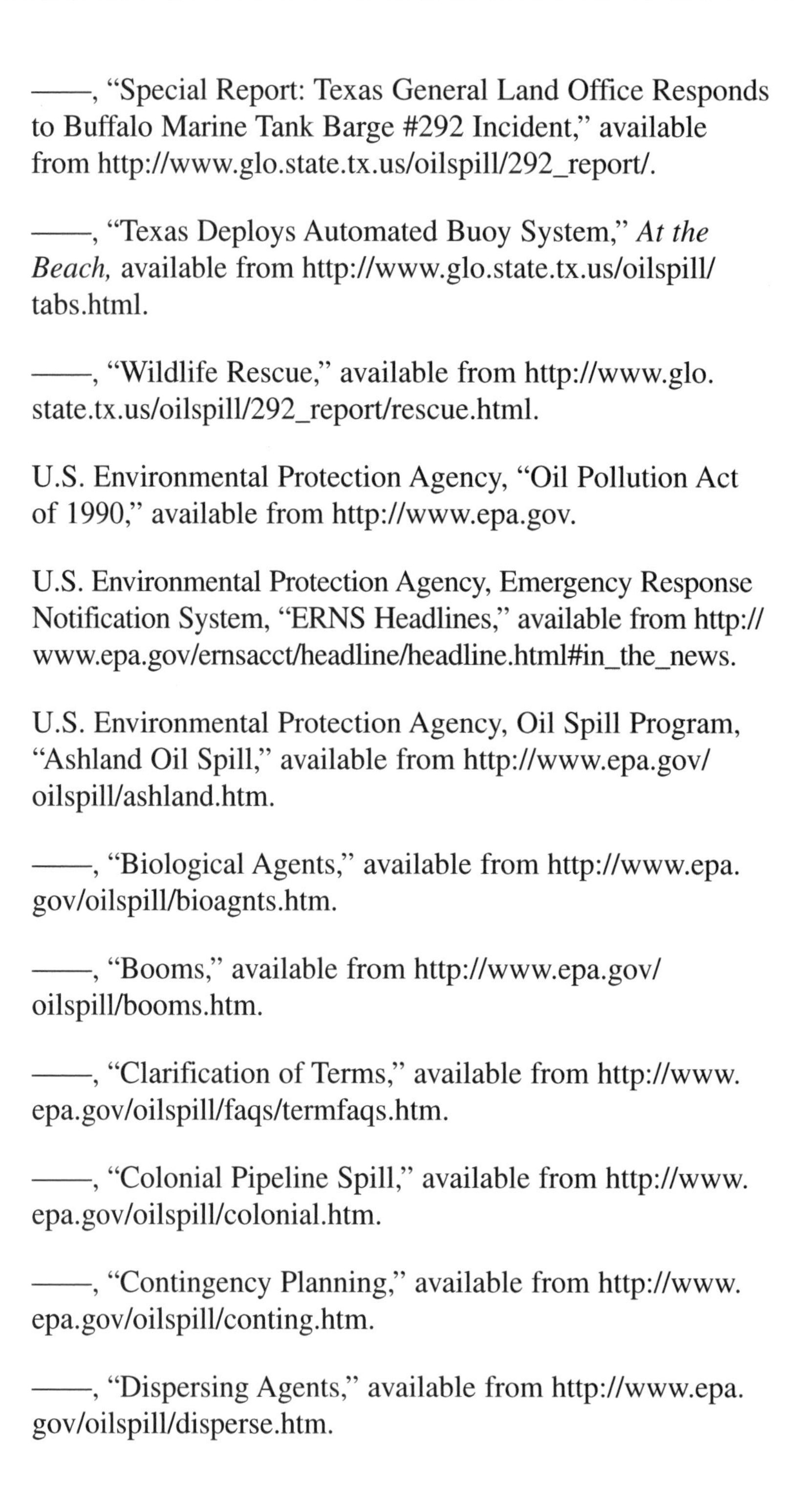

——, "Special Report: Texas General Land Office Responds to Buffalo Marine Tank Barge #292 Incident," available from http://www.glo.state.tx.us/oilspill/292_report/.

——, "Texas Deploys Automated Buoy System," *At the Beach,* available from http://www.glo.state.tx.us/oilspill/tabs.html.

——, "Wildlife Rescue," available from http://www.glo.state.tx.us/oilspill/292_report/rescue.html.

U.S. Environmental Protection Agency, "Oil Pollution Act of 1990," available from http://www.epa.gov.

U.S. Environmental Protection Agency, Emergency Response Notification System, "ERNS Headlines," available from http://www.epa.gov/ernsacct/headline/headline.html#in_the_news.

U.S. Environmental Protection Agency, Oil Spill Program, "Ashland Oil Spill," available from http://www.epa.gov/oilspill/ashland.htm.

——, "Biological Agents," available from http://www.epa.gov/oilspill/bioagnts.htm.

——, "Booms," available from http://www.epa.gov/oilspill/booms.htm.

——, "Clarification of Terms," available from http://www.epa.gov/oilspill/faqs/termfaqs.htm.

——, "Colonial Pipeline Spill," available from http://www.epa.gov/oilspill/colonial.htm.

——, "Contingency Planning," available from http://www.epa.gov/oilspill/conting.htm.

——, "Dispersing Agents," available from http://www.epa.gov/oilspill/disperse.htm.

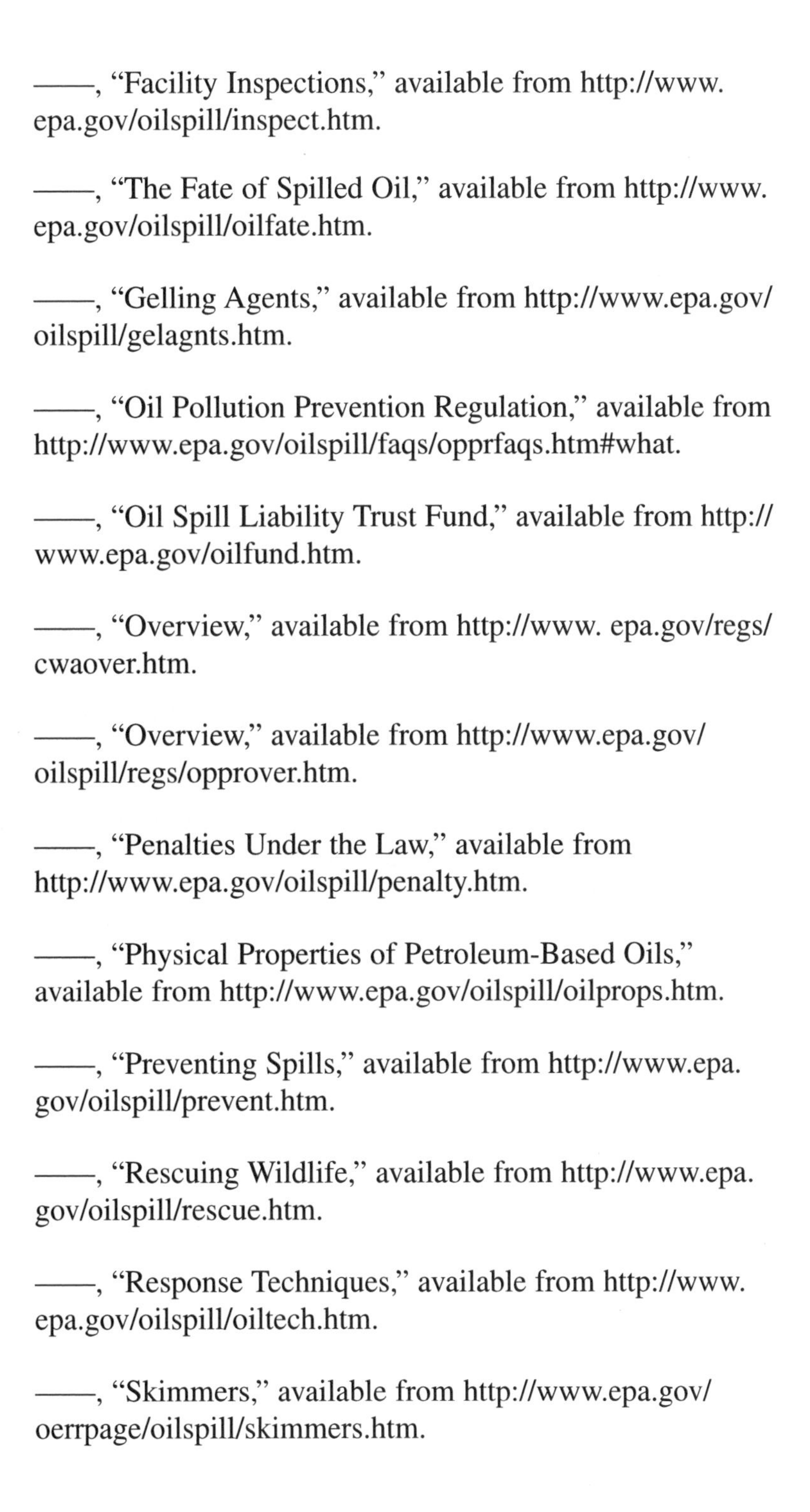

——, “Facility Inspections,” available from http://www.epa.gov/oilspill/inspect.htm.

——, “The Fate of Spilled Oil,” available from http://www.epa.gov/oilspill/oilfate.htm.

——, “Gelling Agents,” available from http://www.epa.gov/oilspill/gelagnts.htm.

——, “Oil Pollution Prevention Regulation,” available from http://www.epa.gov/oilspill/faqs/opprfaqs.htm#what.

——, “Oil Spill Liability Trust Fund,” available from http://www.epa.gov/oilfund.htm.

——, “Overview,” available from http://www. epa.gov/regs/cwaover.htm.

——, “Overview,” available from http://www.epa.gov/oilspill/regs/opprover.htm.

——, “Penalties Under the Law,” available from http://www.epa.gov/oilspill/penalty.htm.

——, “Physical Properties of Petroleum-Based Oils,” available from http://www.epa.gov/oilspill/oilprops.htm.

——, “Preventing Spills,” available from http://www.epa.gov/oilspill/prevent.htm.

——, “Rescuing Wildlife,” available from http://www.epa.gov/oilspill/rescue.htm.

——, “Response Techniques,” available from http://www.epa.gov/oilspill/oiltech.htm.

——, “Skimmers,” available from http://www.epa.gov/oerrpage/oilspill/skimmers.htm.

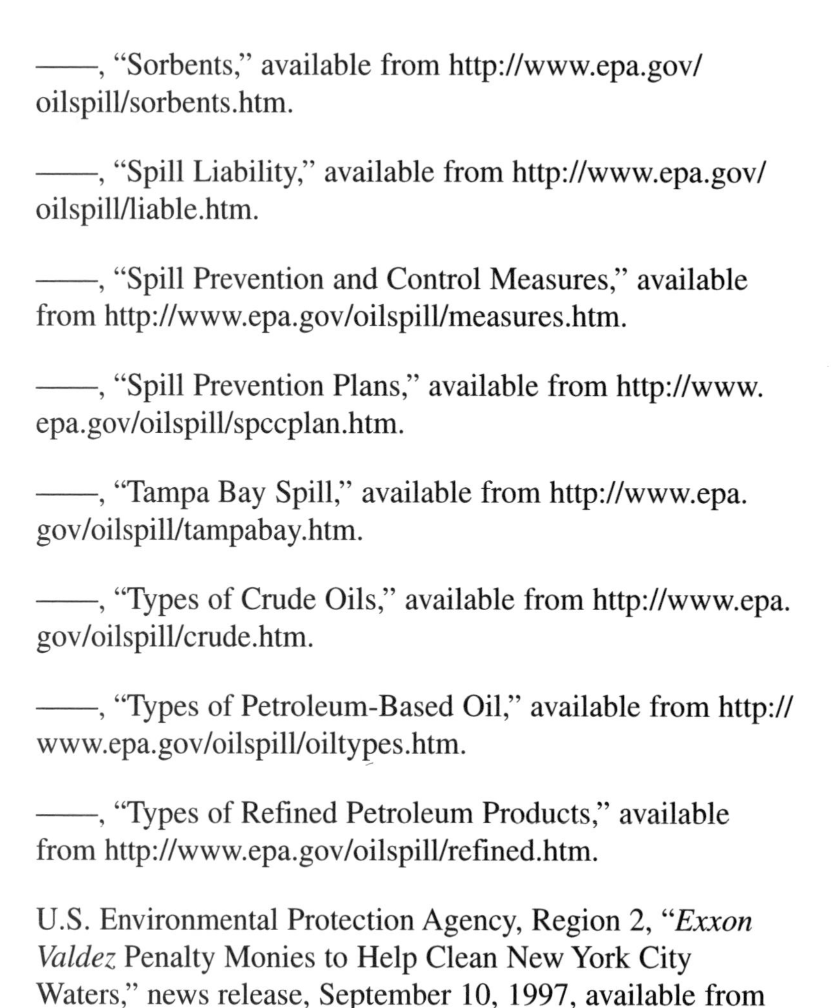

——, "Sorbents," available from http://www.epa.gov/oilspill/sorbents.htm.

——, "Spill Liability," available from http://www.epa.gov/oilspill/liable.htm.

——, "Spill Prevention and Control Measures," available from http://www.epa.gov/oilspill/measures.htm.

——, "Spill Prevention Plans," available from http://www.epa.gov/oilspill/spccplan.htm.

——, "Tampa Bay Spill," available from http://www.epa.gov/oilspill/tampabay.htm.

——, "Types of Crude Oils," available from http://www.epa.gov/oilspill/crude.htm.

——, "Types of Petroleum-Based Oil," available from http://www.epa.gov/oilspill/oiltypes.htm.

——, "Types of Refined Petroleum Products," available from http://www.epa.gov/oilspill/refined.htm.

U.S. Environmental Protection Agency, Region 2, "*Exxon Valdez* Penalty Monies to Help Clean New York City Waters," news release, September 10, 1997, available from http://www.epa.gov/r02earth/epd/exxonnyc.htm.

Index

Picture Credits

Cover photo: © Tony Stone Images/Ken Graham
Alaska Division of Tourism, 9, 65
© Mark C. Burnett/Photo Researchers, Inc., 16
Digital Vision, 15, 31, 51, 53
International Bird Rescue Research Center, 52, 60
Lineworks, Inc., 7
Jim Mayhew/Corbis-Bettmann, 10
© Richard Newman, 20
NOAA, 23
Reuters/Mike Blake/Archive Photos, 44
Reuters/Jim Bourg/Archive Photos, 38
Reuters/Richard Ellis/Archive Photos, 18
Reuters/Eriko Sugita/Archive Photos, 36, 48
UPI/Corbis-Bettmann, 25, 42
U.S. Fish and Wildlife Service, 35, 54, 57

About the Author

Lesley A. DuTemple is the author of many natural history books for children and young adults, covering such topics as tigers, whales, polar bears, moose, and others. DuTemple lives on the edge of a canyon in Salt Lake City, Utah. She, her husband, and two young children share their property with a family of raccoons, a resident porcupine, several flocks of quail and songbirds, two peregrine falcons, and roaming herds of deer.